Field Guide to
Birds
of Britain and Europe

> ➤ **Up to the size of a sparrow**
>   **Pages 26 – 67**

This section contains small songbirds, for example tits, swallows, warblers, robins, finches and sparrows.
(Picture: Tree Sparrow)

> ➤ **Up to the size of a blackbird**
>   **Pages 68 – 105**

This section contains birds from the size of a sparrow to that of a blackbird, including larks, smaller species of thrush and quail, and small woodpeckers.
(Picture: Blackbird)

> ➤ **Up to the size of a pigeon**
>   **Pages 106 – 131**

This section contains birds larger than a blackbird and up to the size of a common pigeon, e.g. lapwing, cuckoo, jackdaw and partridge.
(Picture: Domestic Pigeons)

For ease of identification, birds have been divided into five colour groups.

> **Up to the size of a crow**
> **Pages 132 – 155**

This section contains birds up to the size of a Carrion Crow, for example gulls, crows, magpies, owls and the smaller species of duck. (Picture: Carrion Crow)

> **Larger than a crow**
> **Pages 156 – 185**

In this section you will find birds of prey, the Stork, Eagle-owl, Raven, Capercaillie, as well as the larger ducks, geese and swans. (Picture: Goshawk)

LENGTHS:

From the tip of the beak to the tip of the tail

 Length up to 15cm

 15 to 25cm

 25 to 35cm

 35 to 47cm

 over 47cm

Step-by-step identification

Pied Flycatcher
Ficedula hypoleuca

DESCRIPTION Roughly 13cm long; male breeding plumage ① ranges from dark, grey-brown to deep black, with white underside, forehead and wing patches; autumn plumage similar to female ④, i.e. less contrasting plumage and dark forehead. wing patches of both male and female can be easily observed in flight ③. Juvenile has dense speckling ②.

VOICE: Call is a sh— d clear 'whitt' or a sharp 'tic'; melanc— consists of a rising and falling series o—

DISTRIBUTION: In tall woodland and in parks and gardens with established trees; very common in certain areas; migrant, wintering in tropical Africa.

FOOD: Insects caught either in flight, or picked from branches and leaves; in late summer and autumn also eats berries.

NESTING: Nests in holes in trees, and also in nesting boxes; one–two clutches of five–seven pale turquoise eggs.

④

TYPICAL FEATURES
The Pied Flycatcher has a characteristic wing flutter after landin— —es the Spott— —her.

③

⑤

46

47

Step 1: COMPARE WITH THE MAIN PICTURE

Each main picture shows the bird in its natural habitat with its distinguishing features. Easily confused species are often shown on the same page.

Step 2: IDENTIFICATION OF DISTINGUISHING FEATURES

The two illustrations and additional photograph clearly demonstrate the distinguishing features and provide useful additional information for identification. Using these pictures will enable you to identify a species with certainty.

Step 3: Identification text

The identification text clearly describes important classification features, including detailed indications of bird sizes. Bird calls and songs are also described, along with the bird's preferred sources of food. In addition, there is information about habitat, and whether the bird is native to the United Kingdom or is a rare or occasional visitor. This can often be a big help in identifying birds. A very few of the birds in this book have never been seen in the United Kingdom. Finally, the identification text also gives tips on how to distinguish a bird from other similar species.

Step 4: Calendar clock

Time of year is also helpful for identifying species of bird. Months in which the species can be found in Europe are shaded light grey. The dark grey segments indicate the incubation and nesting periods, i.e. the period from the laying of the first egg, to the time when the nestling is no longer dependent. This is just a guide, of course, and exceptions are possible depending on the climate.

Bird can be found in Central Europe

Incubation and nestling period

Step 5: Infobox

The coloured Infobox offers important additional information about distinguishing or typical behaviour. Together, these steps can help you to identify accurately the desired species.

TYPICAL FEATURES

The Pied Flycatcher has a characteristic wing flutter after landing, as does the Spotted Flycatcher.

Birds

> Cottage gardens are a favourite habitat of the wren.

The fascinating world of birds

Birds have always fascinated people, who have marvelled at their ability to fly, their bright, multi- coloured plumage and their beautiful song.

Europe is home to an enormous variety of birds, from the very small species, such as the Wren and Firecrest, which measure less than 10cm from beak to tail, to the majestic storks, swans and eagles, which can have a wingspan of over two metres.

In fields, woods and meadows

Most species of bird have adapted to a particular habitat. This means that the surroundings in which you spot a bird can offer useful clues as to the species. A mountain bird, such as the Golden Eagle, will never be

> Long-eared Owl (juvenile)

found on the plains of southern England, and birds that like the open countryside, such as the Skylark, will never be seen in deep forests. Birds know how to use the characteristics of their chosen habitat to their advantage. That is why many birds of prey nest in the shelter of woodland, but hunt for their prey in open country.

Man's neighbour

Some native species of bird, for example the House Sparrow, the Swallow and the House Martin, can now be found only in villages, towns and cities. They have become almost like human stalkers, and profit from the presence of people, using our

> Osprey

buildings as shelter and nesting grounds, and feeding exclusively from rubbish tips and titbits left by people, rather than searching for new sources of food.

> Song Thrush

> Grey Partridge

WHY LATIN NAMES?

Each species of bird has a Latin or scientific name in addition to its English name. Some birds may have different common names in different English-speaking countries and even in different regions of the same country, but the Latin or scientific name, consisting of two words, is used internationally to denote the same species. The first word, the genus, indicates closely related families of birds; the second word is the specific name of the species and is always written in lower case letters.

9

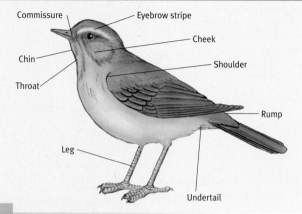

Diagram of body parts and feather groups.

Labels: Commissure, Eyebrow stripe, Cheek, Chin, Shoulder, Throat, Rump, Leg, Undertail

Fancy feathers

Birds are the only animals with feathers. Apart from the beak and legs, a bird's body is completely covered in them. The plumage gives the bird's body its smooth and compact shape, it keeps the bird warm and protects it from sunburn and the rain. Finally, a bird's plumage also provides each species with its distinctive shape and colouring. Nearest to the body is a layer of soft down, and the top layer of plumage comprises glossy contour feathers. The leading edges of the wings mainly comprise sturdy primary feathers.

Juvenile plumage

Nestlings and fledglings are entirely covered in a layer of insulating down plumage. The first covering of true contour feathers, which enables the young bird to fly, is called the juvenile plumage and is usually a different colour or in a different arrangement to the

DISTINCTIVE WING MARKINGS

Many birds have markings in contrasting colours on their wings. Ornithologists use various expressions for these, depending on their shape and position. A wing bar, as seen for example on a Great Tit, is a stripe on the shoulder part of the wing. A wing stripe, for example on a Goldfinch, extends over the base of the primary feathers. Ducks often have wing stripes along the wing contours.

adult plumage. Some species, such as gulls and eagles, only achieve their full adult colour and plumage after a couple of years.

Moulting

All birds preen their feathers regularly, but even with this careful maintenance plumage eventually becomes worn and can no longer adequately perform its functions. Plumage must therefore be regularly renewed. This renewal, called a moult, can take place, depending on the species, either all in one go or in stages, with some birds moulting as many as three times a year, and others only every two or three years. Some species develop a particularly colourful or contrasting plumage before the breeding season, returning to a more simple, inconspicuous plumage once breeding is over.

Wing bar on a
Great Tit

Wing stripe
on a Goldfinch

Wing markings
of a Garganey

> Wing bar on a Great Tit.

> A Grey Wagtail with its white eyebrow stripe.

> Goldfinch with a yellow wing stripe.

> Striped chest feathers of a Goshawk.

> Flecked chest feathers of a juvenile Goshawk.

> Marsh Harrier (male)

> Common Buzzard

Birds in flight

Birds often have a distinctive silhouette in flight, or show characteristic flight patterns. For example, many finches and the Great Spotted Woodpecker can easily be identified by their undulating flight pattern, characterised by a series of quick wing beats, followed by a short gliding flight. Some birds fly in a straight line, while others have a more meandering flight. Some are very agile, others are more ponderous, and some birds have a very shallow wing flap, while others have a very exaggerated wing movement.

Watching the skies

Birds of prey, in particular, are often seen up in the sky, gliding and circling on air currents. Plumage characteristics are often not visible at this distance, and these birds are often only identifiable by their silhouettes.

> Red Kite

> Golden Eagle (Juvenile)

> White Stork

> Common Kestrel

> Mute Swan

> Short-eared Owl

> Grey Heron

The beak:
a multi-purpose tool

The bird beak, often also called the bill, is made out of a hard, rigid material, and is a specially designed, multi-purpose tool. It can be used for seeking, gripping and shredding food; and it is ideal for transporting nesting material and carefully preening feathers. When necessary, a beak can also serve as an effective defensive weapon.

Beaks have developed very differently in order to adapt to the habitat and lifestyles of their owners. Beaks can therefore not only help identify species of bird and closely related families, but the shape and size of a beak can also indicate their food and feeding habits.

Made-to-measure beaks

Feeding habits can vary greatly between different species of bird, and not every beak is suitably adapted to every type of food or to every application. Birds that obtain their food from shallow, muddy waters need long bills, but a duck's bill would be no use for a bird that needs to tear a freshly caught mouse into manageable pieces, and birds that feed on seed kernels need a beak suitable for cracking open shells.

LONG BILLS

Many wading birds, such as the Oystercatcher (pictured), Heron, Stork, Woodcock and Blacktailed Godwit, have a beak that is very long and thin in comparison to their head. This enables the bird to poke in the sand and mud at low tide for lugworms, snails and crabs.

MEDIUM LENGTH, POWERFUL BEAKS

A medium-length beak can be used both to kill prey and to break up hard plant-based foods. Gulls use their beaks to shred fish, crows to rip up carrion and eat crops, Jays (pictured) to open acorns and woodpeckers to hammer and chisel into tree trunks.

SMALL, SLENDER BEAKS

Slender, pincer-like beaks, such as those of tits and warblers (pictured: Garden Warbler), are the perfect tool for picking up insects and their larvae from branches and digging them out of holes. They are also ideal for diets of soft berries.

SHORT, POWERFUL BEAKS

A particularly thick and powerful beak, like that of a sparrow or a finch (pictured: Brambling), is needed to break open hard-shelled seeds and to crush them between the two beak halves. Hawfinches are even capable of cracking open cherry stones and plum stones.

HOOKED BEAKS

Birds of prey and owls (pictured: juvenile Long-eared Owl) have a sharp, hooked upper beak, which is perfectly adapted for catching, holding and tearing prey into small, beak-sized chunks that can then be easily consumed.

DUCK AND GOOSE BILLS

The wide bills have two rows of tooth-like serrations around the edges. In ducks, this serves as a kind of sieve for filtering food particles from the water. In geese (pictured: Bean Goose), it enables them to graze on grasses.

> The Yellowhammer can even be heard singing in the middle of the day.

A chorus of song

Many birds have a very character-istic song and several species are easily identifiable by sound alone. One obvious example is the Cuckoo, which even young children can recognise. Birdsong can be a useful identification tool, even for the most experienced ornithol-ogist. For example, the Willow Tit and the Marsh Tit, and the Willow Warbler and Common Chiffchaff, have a very similar appearance and in the wild are often only distin-guishable by their song.

A song is like a flag

Birdsong, especially in spring and in the countryside, fills the morning and evening air. The singing or call is often interpreted as a sign of simple joy or contentment, but it fulfils a serious function as well. Birdsong is used to indicate to other members of

Nightingale	Marsh Warbler	Common Redstart	Robin
night		4.00 a.m.	4.30 a.m.

the same species that a territory is occupied. It fulfils the same function as a flag waving over a castle. The singer, usually male, uses his song to warn rivals that they have crossed into his territory, as well as to inform females that a partner is available. Finally, the song of both partners can come together in a courtship and mating call.

The higher, the better

A flag can be seen from further afield if it is hung high up and allowed to unfurl freely. The same is true for birdsong, and mating and territorial calls travel further, the higher the bird's perch. For this reason, many songbirds release their songs from the tops of trees and telegraph poles, from rooftops or from any elevated location. For some, even this is not high enough, and they sing while in flight.

Not only songbirds sing

Almost half of all bird species are songbirds, but not all songbirds are talented singers. Due to certain anatomical characteristics, for example the shape of the larynx and feet, and other behavioural similarities, crows are classified by zoologists as songbirds. So, despite what the name suggests, songbirds are not only so-called because of the delightful sounds they make! In terms of function, the 'laugh' or 'yaffle' of a Green Woodpecker, or the throaty 'og-og-og' of a Woodcock are actually birdsong.

On the other hand, some birds that are not officially classed as songbirds can 'sing' beautifully.

> The Starling – the prima donna of the songbirds.

Chaffinch	Wren	Blue Tit	Yellowhammer
5.00 a.m.	5.30 a.m.	6.00 a.m.	midday

17

Birds and migration

Flight has enabled birds to inhabit even the most northerly climes, where food is not available all year round. During the cold season, when food can become scarce, birds can simply fly away to regions with a better climate. The exact timing of their migration, and the regions they migrate to, vary between the species, depending on their particular feeding habits and climate preferences. Slight fluctuations in the timing of migration flights are also possible from year to year. Some species, especially long-distance migrant birds, such as storks, seem to have an in-built timer, and always migrate at exactly the same time, taking exactly the same route to their winter destination.

Different ways of travelling

Some migrant birds travel alone or in small groups. Others migrate in enormous flocks. Geese typically travel in a chain or in a v-formation. Some species, for example Terns, travel thousands of miles almost without resting, others make their journey in a series of short hops over a longer period of time. In order to avoid dangers and predators, many migrant birds, for example the Redwing, undertake their long flights under cover of darkness, and use the daylight hours to rest.

House Sparrow

PERMANENT RESIDENTS

Permanent residents are species of bird that can be found all year round in their breeding and nesting grounds. Some species, for example the Nuthatch, never stray from their breeding territory, while others, for example the Coal Tit, roam over wider areas.

Blackbird

PARTIAL MIGRANTS

Partial migrants are species of bird of which some individuals migrate in autumn, while others remain in their breeding ground year round. Many species of bird that winter in parts of Europe simply move south from northern breeding grounds.

Common Redstart

MIGRANT BIRDS

Migrant birds often breed in Europe but migrate to warmer regions during the European winter. Short-distance migrants only travel as far as the Mediterranean or western Europe, but many long-distance migrants travel to the more tropical climate of Africa.

Brambling

WINTER VISITORS

A winter visitor is a bird whose breeding ground is located in more northerly or north-eastern regions, but who winters in Central Europe. Birds whose wintering grounds lie further south and that are only seen at migration time are called passage migrants.

Long-distance fliers

Many migrant birds, particularly the long-distance fliers, perform amazing feats of endurance. The Aquatic Warbler, for example, flies from its central European breeding ground all the way to the tropics of West Africa, covering the 2,700 miles in just three or four days. Garden Warblers also cross the Mediterranean and the Sahara without stopping as they migrate south. To prepare for these endurance flights, the birds feed extensively to build up a layer of excess fat, and they can put on as much as thirty per cent of their normal body weight.

Should you leave food out over winter?

The cold weather presents a problem for those birds that choose to spend the winter in Europe, and it can be a hard time. Many bird lovers try to help the birds survive this tough season by putting food out on bird tables. The advantages of winter feeding are, however, questioned by many bird conservationists. Bird tables provide nature lovers with a wonderful spectacle, but they do not make any contribution to species protection. In fact, this feeding only helps those species that are already the most common. In order to prevent harm to the birds, for example through the spread of infectious diseases via droppings, bird tables must be regularly cleaned. Birds should never be fed outside winter.

19

> The mating ritual of the Capercaillie is an optical and acoustic spectacle.

Courtship behaviour and nest building

The primary aim of the courtship ritual is for the male birds to attract females using a variety of techniques. The exact techniques differ from species to species but can include elaborate flight displays (for example, the Lapwing) or displays of size and beauty, as demonstrated by the Capercaillie.

A tailor-made nest

When a male and a female come together, both partners perform a series of highly ritualised movements, reminiscent of a kind a dance, in order to consolidate the partnership.

The location and design of a nest, and the materials from which it is constructed, are very species-dependent. The photos on the opposite page show just a small sample of the numerous styles and types of nest.

LESS CURIOSITY AND MORE CONSERVATION

Bird-lovers should take care not to disturb breeding pairs and nests, not only in designated bird conservation areas, but in all nesting sites.

> Ball-shaped nest of a Chiffchaff.

> Reed nest of the Little Grebe.

> Wren's nest constructed from moss and leaves.

> Osprey nest

> Suspended nest of the Marsh Warbler.

21

> Hungry Yellowhammer chicks eager for a feed.

Nestlings

The number of eggs laid by the female bird per clutch varies between the species. Many of the larger birds of prey only lay one or two eggs, but other birds, such as the Grey Partridge, regularly lay between 10 and 20 eggs.

A brief glimpse of the eggs in the clutch is often sufficient to identify the species. The size, shape, colour and patterning of the eggs are different for each species. The photos on the opposite page are a small example of the great variety of bird's eggs.

Nidicolous birds

There are two categories of chick. The first category, that of nidicolous birds, are hatched in a very undeveloped state. They are initially bald and blind, and only

their digestive system is in perfect working order. As soon as they feel the slightest movement of the nest, the chicks begin to call. They raise their wide-open beaks and

AN ORPHANED CHICK?

The young of many species of bird leave the nest before they are able to fly. They sit on branches or on the ground and are fed there by their parents. If you find a young bird like this, you should not take the little chick home. The bird is almost certainly not an orphan. Rearing chicks requires a lot of experience, and amateurs are rarely successful. The chick has a much greater chance of survival if left alone.

wait for food to be dropped into their mouths. While their plumage remains unformed, the chicks must must be kept warm beneath the wing or breast feathers of an adult bird. This is called brooding.

Pigeons, birds of prey, owls, woodpeckers and songbirds are typically nidicolous.

Nidifugous birds

In contrast, nidifugous birds are sufficiently developed upon hatching to be able to leave the nest within a couple of hours, or at the latest within a couple of days. Accompanied by one or both adult birds, the chicks are then able to search for food more or less independently.

Ducks and geese, chickens, terns and wading birds are typical nidifugous birds.

› Icterine Warbler eggs

› Song Thrush eggs

› Mallard eggs

› Blackbird eggs

› Young Song Thrush

› Sparrow Hawk eggs

23

List of birds

Firecrest
Regulus ignicapillus

DESCRIPTION: This small, round songbird measures just 9cm from the tip of the beak to the tip of the tail. It has a prominent stripe on its crown, yellow in females ③ and more orange in males ①, but always with a black border. It also has a black eye-stripe and a clearly defined white eyebrow-stripe ④.

VOICE: High 'see-see' call; song is a high and quick 'see-see-see-see-tsit'.

DISTRIBUTION: Widespread breeding bird found in coniferous and mixed woodland, and even in cemeteries, parks and gardens; migrant bird, spends the winter in the Mediterranean.

TYPICAL FEATURES
The Firecrest can be seen quite frequently 'hovering' in the air between two branches while looking for insects on the branch surfaces.

FOOD: Small insects and spiders.

NESTING: Nest is a deep, thick-walled bowl ②, located in the crook of a branch, or supported by two close branches; two clutches of between seven and 12 eggs per year, with small brown speckles.

Goldcrest
Regulus regulus

DESCRIPTION: Same size and shape as the Firecrest, but with less colourful plumage. The crest on the female is yellow-black ④, while the male has more orange-coloured feathers ①. Juveniles have no crest.

VOICE: High, squeaky 'see-see' call; territorial song is a very high-pitched, fluctuating twitter.

DISTRIBUTION: Widespread and common breeding bird, most often found in spruce woods, but also in cemeteries, parks and large gardens with conifers. Breeding grounds are in

TYPICAL FEATURES
The Goldcrest does not have the black eye-stripe and white eyebrow stripe that are typical of the Firecrest ③.

Central Europe, where it remains all year round. In winter, greater numbers can be found, as some migrate to Central Europe from Scandinavia.

FOOD: Small insects and spiders.

NESTING: Nest is a deep bowl, either in the fork of a branch or suspended between two conifer branches, built entirely by the female. Two clutches of between eight and 10 eggs per year, covered in small brownish speckles.

27

Wren
Troglodytes troglodytes

DESCRIPTION: The Wren measures just under 10cm from the tip of the beak to the tip of the tail. The brown plumage on its wings and flanks displays faint stripes ③, the throat is whitish, and there is a thin, white eyebrow-stripe ②. Its short tail is usually cocked ①.

VOICE: The alarm call is a humming 'zerrrr' or a harsh 'tit-tit-tit'. Warbling song mixed with trills can also be heard in winter.

DISTRIBUTION: Coniferous and mixed forests, lowland forests with plenty of undergrowth, parks and gardens with thick bushes. Most Wrens stay in Europe year round, but the population increases in winter as some birds migrate from their more north-easterly breeding grounds.

FOOD: Small insects and their larvae; spiders and small worms.

NESTING: Two clutches per year of between five and seven reddish speckled eggs. The nest is pictured on p21.

TYPICAL FEATURES

This small bird normally looks for food at ground level, where it scurries through the undergrowth like a mouse. Darts quickly when disturbed.

Dunnock
Prunella modularis

DESCRIPTION: Just over 14cm long and similar to a House Sparrow (p58), but slimmer with a thinner beak ①; breast and head lead-grey, only the crown and cheeks are brown ②; juvenile bird ③ has a whitish throat.

VOICE: Call is a high, slightly croaky 'tsit' or 'di-di'; song is light and warbling, with a slight variation in tone, reminiscent of a creaking door. Can be heard as early as March.

DISTRIBUTION: Mixed and conifer woodland with dense undergrowth; parks and gardens with sufficient undergrowth.

TYPICAL FEATURES

The Dunnock hops on the ground in a very crouched stance. Darting flight, very close to ground.

FOOD: Insects; in autumn and winter also eats seeds and berries.

NESTING: Bowl-shaped nest constructed from grass stalks and moss, not too far above the ground, located in dense undergrowth or in spruce forests. Two clutches of four–five turquoise eggs per year.

SIMILAR SPECIES: The slightly larger Alpine Accentor *(Prunella collaris)* ④ is normally only found at altitudes of between 1,500 and 2,300m. It has red-brown speckled flanks and a pure grey head.

29

Tree Pipit
Anthus trivialis

DESCRIPTION: This slim songbird is about 15cm long and its brown speckling is unremarkable ①. Its belly can vary in colour from creamy white to a more yellow-brown shade and it has dark, broken stripes on the breast ②.

VOICE: Call is a high, hoarse 'tseea' or an urgent 'tsip-tsip'; melodic song similar to a Canary.

DISTRIBUTION: In open countryside with occasional clusters of trees; woodland edges and clearings; winters in Africa.

TYPICAL FEATURES
In song flight, the Tree Pipit climbs steeply from its perch, then glides back down with outstretched wings and spread tail while producing his song ③.

FOOD: Mainly insects, but also spiders and other small animals.

NESTING: Nests on ground, usually in tall grass; normally two clutches of four–six eggs per year; variable colouring but always densely speckled.

SIMILAR SPECIES: The Water Pipit *(Anthus spinoletta)* ④, which breeds in mountainous areas above the treeline, has a more grey colouring.

Meadow Pipit
Anthus pratensis

DESCRIPTION: Slightly smaller than the Tree Pipit and has a more grey or olive-brown coloured back ①; streaked breast; after moulting in spring, its belly is ochre ②; juvenile ③ has fewer streaks.

VOICE: During ascent and in flight, often utters a clear, light 'heet' call; song consists of monotone phrases and a series of liquid notes; sings in flight, occasionally also from trees ④.

TYPICAL FEATURES
The Meadow Piper normally begins its song flight on the ground, but, at the end of this flight, it lands in a different spot.

DISTRIBUTION: Damp fields and meadows, moorland and heathland. Common breeding bird in lowlands of Northern Europe, less common in southern parts. Some spend the winter in the Mediterranean, others, mainly migrants from more north-easterly breeding grounds, spend the winter in western Europe, including England and Wales.

FOOD: Almost exclusively insects and spiders; feeds on the ground.

NESTING: Nests on the ground, well-hidden in vegetation; one–two clutches of four–six eggs per year; densely speckled with grey or red-brown.

31

Chiffchaff

Phylloscopus collybita

DESCRIPTION: This songbird is about 11cm long; inconspicuous with olive-brown plumage, which is lighter on the belly ①; its yellow eyebrow-stripe ④ is sometimes hard to distinguish; legs are normally almost black ③.
VOICE: Call is a monosyllabic 'tsip'; monotonous song consists of a long, stammering 'chiff-chaff-chiff-chaff', interspersed with a soft chirping call.

> **TYPICAL FEATURES**
> The Chiffchaff can often only be told apart from the Willow Warbler by its song.

DISTRIBUTION: In light deciduous and mixed woodland, lowland forests and field coppices; also in parks and gardens; spends winter in the Mediterranean.
FOOD: Small insects, spiders, also berries.
NESTING: Nest with side entrance (p21), hidden in vegetation close to ground level; one–two clutches of five–six brown speckled eggs per year.
SIMILAR SPECIES: The Dusky Warbler *(Phylloscopus bonelli)* ② breeds in forests and has more yellow in its plumage. It is rare.

Willow Warbler

Phylloscopus trochilus

DESCRIPTION: The Willow Warbler is very similar in appearance to the Chiffchaff, but its plumage is often somewhat yellower ①. Its yellow eyebrow-stripe is usually more prominent and its legs are a light flesh colour ②. Juveniles ③ often have a distinctive yellow belly.
VOICE: Melancholy and wistful song consists of a descending series of soft, flute-like tones.
DISTRIBUTION: In light woodland of any type, moors, lowland forests, gardens and parkland,

> **TYPICAL FEATURES**
> The Willow Warbler's call is similar to that of the Chiffchaff but has two distinct syllables, a bit like a 'tsi-ip'.

preferably with birches and meadows; winters in tropical Africa.
FOOD: Small insects, spiders and other small animals that it catches while climbing around in the branches; in autumn also eats berries.
NESTING: Nest is a covered construction with a side entrance, hidden in low vegetation; one–two clutches of four–seven red-speckled eggs per year.
SIMILAR SPECIES: The Wood Warbler *(Phylloscopus sibilatrix)* ④, lives in beech woods, and has a bright sulphur-yellow-coloured throat.

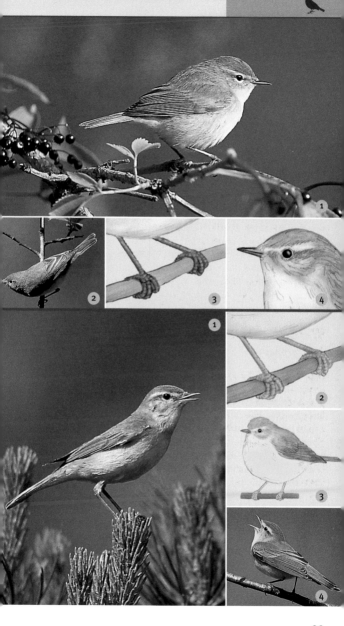

33

Marsh Warbler

Acrocephalus palustris

DESCRIPTION: This slim, 12cm-long songbird has a brown back, often with a slight olive tone ①; rump is a pale red-brown ②; face has a narrow, beige eyebrow-stripe ③.

VOICE: Alarm call is a harsh 'tchak'; song is pleasant, but very varied, as the Marsh Warbler imitates the songs of other species.

TYPICAL FEATURES
This tireless singer usually performs its vast repertoire from within deep cover, even when it is dark.

DISTRIBUTION: Common in lowlands and found in many other areas, but it is hard to spot; lives in thickets, for example in clumps of willow and stinging nettles; often, but by no means exclusively, found in marshy areas; migrates to tropical Africa for the winter.

FOOD: Small insects.

NESTING: Deep, bowl-shaped nest, hanging from closely packed plant stalks ④; one clutch of four–five eggs per year; eggs have olive-green speckles.

Reed Warbler

Acrocephalus scirpaceus

DESCRIPTION: The Reed Warbler is very similar to the Marsh Warbler in appearance; its back is however more of a red-brown colour, with a prominent rust-red rump ②.

VOICE: Call is a short and inconspicuous 'tchuk' or 'tch', but a rasping 'fit' when alarmed; song is incessant, somewhat hasty and monotonous, with a rhythmic motif: 'tiri-tiri-tiri-shirk-shirk-shirk-tserr-tserr'.

TYPICAL FEATURES
The Reed Warbler can often be seen flying close to the reeds, almost skimming their tops.

DISTRIBUTION: Very common in lowlands; breeds only in reeds; only found away from water when in passage; winters in tropical Africa.

FOOD: Small insects.

NESTING: Deep, sturdy, bowl-shaped nest, hanging between two reed stalks ①; three–five pale green eggs with dense, dark, grey-green speckles.

SIMILAR SPECIES: The Sedge Warbler (*Acrocephalus schoenobaenus*) ④ also inhabits reed beds, but it has a prominent white eyebrow-stripe ③.

35

Icterine Warbler
Hippolais icterina

DESCRIPTION: This songbird is about 13cm long, with olive-grey colouring on its back; the belly is bright or dull yellow ②. Its legs are dark grey, almost black, and the tip of the tail is cut straight across ③.

VOICE: Alarm call is a noisy 'ts-ts-ts', a three-syllable 'te-te-roid' or a four-syllable 'dje-dje-dje-liu'; song is loud and very varied; frequently imitates the songs of other species.

TYPICAL FEATURES

In the mating season, the Icterine Warbler sings all day long, endlessly hopping through the branches ④.

DISTRIBUTION: Very common lowland breeding bird, in light deciduous and lowland woodland, and also in gardens and parks; spends the winter in tropical Africa.

FOOD: Insects and their larvae, spiders and other small animals; in autumn also eats berries.

NESTING: Deep, bowl-shaped nest, normally wedged in the fork of a branch ①; four–six bright pink eggs, with occasional black flecks (p23).

Whinchat
Saxicola rubetra

DESCRIPTION: About 12.5cm long; the breeding plumage of the male (①, left in the picture) has an orange-brown throat and breast, and a dark brown head and back; the winter plumage ③ hardly differs from that of the female (①, right in the picture); short tail and brown-flecked rump ②.

VOICE: When agitated, a harsh 'tik-tik', interspersed with a softer 'tiu'; song consists of short, very varied phrases, with metallic whistling.

TYPICAL FEATURES

Both male and female Whinchats always have a broad, white eyebrow-stripe. The Stonechat lacks this marking.

DISTRIBUTION: In meadows, moors and marshland; rare breeding bird in Central Europe; winters in South-west Europe or North Africa.

FOOD: Small insects and their larvae; spiders.

NESTING: Nests on the ground, well-hidden in tall grass; five–six turquoise-coloured eggs.

SIMILAR SPECIES: The male of the even rarer Stonechat (*Saxicola torquata*) ④ has a black throat; the female is very similar to the Whinchat.

37

Bluethroat
Luscinia svecica

DESCRIPTION: About 14cm long; the male breeding plumage ① has a brilliant blue throat with a white 'spot'; in some birds this spot is red ④; the winter plumage has just a thin, blue breast stripe; female ③ has the same plumage year round, with a whitish, black-edged bib.

VOICE: Call is a harsh 'tack' or a whistling 'huit'; song is chirping and churring, with whistling tones interspersed with imitations of other bird songs.

TYPICAL FEATURES

The Bluethroat often flashes its tail, revealing its rust-red tail coverts. These are also visible during flight ②.

DISTRIBUTION: Breeding bird, normally found in the Alps and occasionally in lowland areas; likes marshy areas and reed-covered banks; winters in the Mediterranean and North Africa.

FOOD: Mainly insects and spiders.

NESTING: Nests in dense vegetation, not far off the ground; five–seven eggs, colour varies greatly from grey-green to rust-coloured.

Robin
Erithacus rubecula

DESCRIPTION: About 14cm long and greenish-brown ①; especially when it is cold, the Robin can look very rounded ③; red breast, neck and face ②; juvenile has brown, heavily speckled plumage, without the red ④.

VOICE: Call is a sharp 'tic' which increases to a quick, chattering 'tic-tic-tic' when alarmed; territorial song is a light, rising warble, normally performed from the cover of bushes or trees; song continues until late dusk.

TYPICAL FEATURES

The Robin can often be seen looking for food on the ground. When doing so, it often twitches and flashes its tail.

DISTRIBUTION: In woods and fields, coppices with dense undergrowth, parks and gardens, even in the middle of cities. In residential areas where there are many bird tables, many Robins will stay for the winter; others spend the winter in the Mediterranean.

FOOD: Insects, spiders, worms; in autumn and winter, also berries.

NESTING: Bowl-shaped nest, constructed from stalks, leaves and moss and concealed close to the ground; normally two clutches of five–seven eggs with variable speckling.

Black Redstart
Phoenicurus ochruros

DESCRIPTION: About 14cm long; both male and female have rust-red tail and rump; male ① is usually ash-grey and black, with prominent white markings on the wing ②; the female ③ is plain, dark, grey-brown.

VOICE: Rapid, harsh 'huid-tucc-tucc' call, or a toneless 'teckteckteck'; song consists of short, rasping phrases, which begin and end with a light whistling; starts singing at early dawn from a high perch, e.g. a television mast.

TYPICAL FEATURES
A young Black Redstart ④ can be distinguished from a young Common Redstart by its lack of speckling. Both birds have a red tail.

DISTRIBUTION: Originally inhabited rocky mountain regions but has now followed humans and is also common in lowland towns and villages; short-distance migrant.

FOOD: Insects and their larvae, spiders, but also berries.

NESTING: Untidy nests in between rocks, or in wall and roof crevices; normally two clutches of five–six white eggs per year.

Common Redstart
Phoenicurus phoenicurus

DESCRIPTION: About 14cm long, slim with long legs; male ① has a slate-grey back, and a white forehead ③; has a bright, rust-red colour, not only on its tail and rump, but also on its breast; looks very colourful in flight ②; female ④ has brown back, beige belly, with rust-red tail and rump.

VOICE: Harsh 'huid-tucc-tucc' call when alarmed; song begins with a high whistle, followed by some quick, melodious tones; begins singing at dawn or before, from a high perch.

TYPICAL FEATURES
Both the Black Redstart and the Common Redstart give a frequent, quick flick of the tail when perching.

DISTRIBUTION: In light, ancient woodland, rural orchards and town parks and gardens; has become rare in some regions; migrates to Central Africa for the winter.

FOOD: Insects, spiders and berries.

NESTING: Nests in holes in trees or in wall crevices; also known to nest in bird boxes; normally two clutches of five–six pale turquoise eggs.

41

Blackcap
Sylvia atricapilla

DESCRIPTION: This slim songbird is about 14cm long and has an unremarkable brown plumage. The male has a distinctive black cap ① but in the female ④ and juveniles ② this cap is red-brown.

VOICE: When disturbed, a harsh 'tacc-tacc' call, which builds to a screech when highly agitated; pleasant song, which begins with a chattering, twittering sound, followed by rounded, clear, flute-like tones.

DISTRIBUTION: Woods with dense undergrowth; also common in parks and gardens with abundant shrubbery; short-distance migrant.

FOOD: Insects and their larvae, spiders; in autumn also eats berries.

NESTING: Flimsy, shallow, bowl-shaped nest, built in the undergrowth, normally less than 1.5m above the ground. Often two clutches of four–six brown-coloured eggs with dark speckles.

TYPICAL FEATURES
The female is similar to the Common Whitethroat (p44), but this has white borders on the tail. The Blackcap tail ③ has no white.

Garden Warbler
Sylvia borin

DESCRIPTION: 14cm-long, slim songbird, with a uniform grey-brown back ① without distinctive markings. Belly is somewhat lighter ②.

VOICE: Sharp 'fit-fit-fit', when alarmed a repeated, harsh 'tacc'; song is loud and melodious, consisting of long, rounded flute-like phrases.

DISTRIBUTION: Common in woods with dense undergrowth, parks with bushes, and overgrown gardens; migrates to Central or Southern Africa for the winter.

FOOD: Insects and their larvae, spiders; in autumn also eats berries.

TYPICAL FEATURES
This inconspicuous and secretive bird normally performs its song from the cover of thick foliage.

NESTING: Nest found low in thick undergrowth; four–six eggs with a whitish shell and irregular, light brown speckles.

SIMILAR SPECIES: The Barred Warbler (*Sylvia nisoria*) ④ lives mainly on the edges of woods, in bramble undergrowth. It can also be identified by its horizontally banded underside and bright yellow eyes ③.

43

Common Whitethroat
Sylvia communis

DESCRIPTION: Slim warbler, about 14cm long, rusty brown wings ①, pale legs ③ and white edging to the sides of the tail ②; male ① has a grey crown and light pink colouring on the breast; female and juvenile are more brown.

VOICE: Call is a nasal 'foid-foid' or a loud 'tse'; song is a short, rough chattering, with scratchy, slurring tones; sings from a high perch ④, and also often during a song flight.

DISTRIBUTION: In countryside with abundant brambles, road and rail embankments, and parks and gardens with thorny undergrowth; winters in tropical Africa.

TYPICAL FEATURES
The male can be easily spotted by his short song flight. He begins from a high perch and takes off vertically.

FOOD: Insects and their larvae, spiders; in autumn also eats berries.

NESTING: Nest normally constructed low down in thorny bushes; usually two clutches of four–five light grey, finely speckled eggs per year.

Lesser Whitethroat
Sylvia curruca

DESCRIPTION: Very similar to the Common, but the back is more grey and the belly ② is slightly lighter; almost black legs ③. Its dark grey cheeks stand out from the white throat ①.

VOICE: When disturbed a sharp, noisy 'tacc' or 'tieck'; territorial song begins with a quiet, rattling twitter, followed by a loud, monotonous, wooden-sounding rattling.

DISTRIBUTION: In gardens, parks, cemeteries and orchards; in field coppices and spruce woodland; found from lowlands to above the treeline in mountainous regions. Migrates to tropical Africa for the winter.

TYPICAL FEATURES
The Lesser Whitethroat is more secretive than the Common Whitethroat. Does not have a song flight, and instead sings from the cover of branches ④.

FOOD: Insects and their larvae, spiders; in autumn also eats berries.

NESTING: Shallow, flimsy nest in hedgerows and shrubbery, often also in young conifers; normally only one clutch of four–six eggs per year; eggs have light, but often multi-coloured speckling.

45

Spotted Flycatcher
Muscicapa striata

DESCRIPTION: Roughly 14cm long and slim, inconspicuous grey-brown colour ①; light brown flecking on head and pale breast ②; juvenile has dense speckling.

VOICE: Call is a sharp 'pst' or a light 'see'; when threatened a frantic 'tek-tek-tek'; chirping song, somewhat jerky.

DISTRIBUTION: Wood edges, clearings, parks and large gardens; often nests in houses; long-distance migrant.

FOOD: Flying insects.

NESTING: In semi-enclosed holes in trees and

TYPICAL FEATURES
Like all flycatchers, the Spotted Flycatcher perches and watches out for insects that fly past. It then launches into flight to catch the insects.

walls, and also on roof beams; five–seven pale green, brown-speckled eggs.

SIMILAR SPECIES: The Red-breasted Flycatcher *(Ficedula parva)* males ④ have a prominent orange-red throat patch; females and juveniles similar to Spotted Flycatcher, but can be identified by black and white tail markings ③. They are rarely found in Britain.

Pied Flycatcher
Ficedula hypoleuca

DESCRIPTION: Roughly 13cm long; male breeding plumage ① ranges from dark, grey-brown to deep black, with white underside, forehead and wing patches; autumn plumage similar to female ④, i.e. less contrasting plumage and dark forehead. Wing patches of both male and female can be easily observed in flight ③. Juvenile has dense speckling ②.

VOICE: Call is a short and clear 'whitt' or a sharp 'tic'; melancholy song consists of a rising and falling series of notes.

TYPICAL FEATURES
After landing, the Pied Flycatcher gives a characteristic flash of the tail, similar to the Spotted Flycatcher.

DISTRIBUTION: In tall woodland and in parks and gardens with established trees; very common in certain areas; migrant, wintering in tropical Africa.

FOOD: Insects caught either in flight, or picked from branches and leaves; in late summer and autumn also eats berries.

NESTING: Nests in holes in trees, and also in nesting boxes; one–two clutches of five–seven pale turquoise eggs.

47

Common House Martin
Delichon urbica

DESCRIPTION: This sociable songbird is roughly 12.5cm long and has a characteristic forked tail ② and a pure white belly ④ including the throat ③; rump is also white. Adult birds have a shimmering blue-black back; juvenile plumage is dark brown.

VOICE: Soft 'trr-trr', or 'prrrt' call; alarm call is a high, piercing 'tseep'; song consists of loud, short, twittering phrases.

DISTRIBUTION: Where people have settled, mainly in villages, individual farms and on the edges of towns; winters in sub-Saharan Africa.

FOOD: Small, flying insects, caught in flight.

NESTING: Colonies of mud nests on the external walls of buildings, under bridges or on cliffs; two–three clutches of three–five white eggs per year.

TYPICAL FEATURES

The House Martin's neat nest, constructed from clumps of mud ① is shaped like a quarter sphere with a semicircular entrance hole.

Swallow | Barn Swallow
Hirando rustica

DESCRIPTION: Same size and shape as a House Martin, but with long tail tips ②; white belly is in sharp contrast to the glossy blue-black back ④; only the forehead, chin and throat are bright, rust-red ③.

VOICE: Often calls 'tschit-tchit' or 'weet-a-weet'; alarm call is a sharp 'tswit-tswit'; song is a prolonged twittering, with chattering, clear notes; always ends with a characteristic churring sound.

DISTRIBUTION: Nests in large numbers in villages and isolated farms, sometimes also on the outskirts of towns; seen hunting for food above fields, meadows and parks; winters in tropical Africa.

FOOD: Small insects, caught exclusively in flight.

NESTING: Always nests inside buildings, often in livestock sheds or barns; normally two clutches of four–five white eggs with reddish speckles per year.

TYPICAL FEATURES

The Swallow's nest is bowl-shaped and open at the top, constructed from mud and grass ①, often with long stalks hanging from the bottom.

49

Blue Tit
Parus caeruleus

DESCRIPTION: About 12cm long, blue and yellow tit ①; head has a sky-blue crest, bordered with white, and white cheeks ②, the yellow belly, unlike that of the Great Tit, has only a short, black, central stripe ③; pale juvenile plumage ④ has yellow cheeks and grey crest.

VOICE: Frequent, clear 'tsi-tsi-tsi' call; territorial song has light, clear phrases, which sound like 'tsi-tsi-tsi-tsirrrr'.

DISTRIBUTION: In deciduous and mixed woodland (preferably with oak trees), in field coppices, parks and gardens; very common in places.

TYPICAL FEATURES
When mating in spring, the male has a butterfly-like, fluttering flight, and the female begs the male for food by twitching her wings.

FOOD: Small insects and their larvae, spiders and other small animals; occasionally (especially in winter) also seeds and berries.

NESTING: Felt-like nest of moss, animal hair and feathers, built in holes in trees, nesting boxes or holes in walls; seven to 10 white, red-spotted eggs.

Great Tit
Parus major

DESCRIPTION: Tit, about 14cm long, with olive-green back and black head with white cheeks ①; Belly is yellow with black central stripe, which is broad on the male ④, but noticeably thinner and often also shorter on the female ③.

VOICE: Numerous calls, from a rasping 'cherr-r-r-r' to a 'pink' or a light 'tsitsitsi'; song consists of three or four syllables, which are constantly repeated: 'tsi-tsi-be tsi-tsi-be' or 'pee-too pee-too'.

TYPICAL FEATURES
Juvenile Great Tits can be recognised by their pale yellow, instead of white, cheeks, which also lack the black lower border ②.

DISTRIBUTION: Found in any wooded landscape with the exception of pure conifer forests; large numbers also found in parks, cemeteries and gardens, even in city centres.

FOOD: Small insects and their larvae, spiders, seeds; at bird tables they mainly eat suet, nuts and sunflower seeds.

NESTING: Nests in holes in trees, nesting boxes or other holes; most often two clutches of eight to 12 white eggs with red speckles per year.

51

Coal Tit
Parus ater

DESCRIPTION: Just under 12cm-long tit with light, white wings ①, a broad, black bib on the chin ②, and a white, medium-length stripe on its nape ④; also has dual white wing-bars ③; juvenile has yellow colouring on cheeks, nape and breast.

TYPICAL FEATURES
When searching for food, Coal Tits hop about endlessly and somewhat frantically in the tree tops and on the branch tips.

VOICE: While searching for food, often calls a harsh 'tsi', 'psit' or 'si-si-si'; song is a light 'chuuee-chuuee' or 'sitchu-sitchu-sitchu'; can be heard almost year round.

DISTRIBUTION: Mainly in conifer woodland, but also in parks and large gardens with conifer trees.

FOOD: Small insects and spiders, but also conifer seeds; likes carrion and nuts at bird tables.

NESTING: Flimsy nest of moss, animal and plant fibres and spider webs, built in tree holes, hollow stumps and even abandoned mouse holes; two–three clutches of seven to 11 red-speckled eggs per year.

Marsh Tit
Parus palustris

DESCRIPTION: 11–12cm-long, grey-brown tit with a glossy black cap ①, white cheeks and a small black chin fleck ②; no white wing-bars ③.

VOICE: Call like a heavy sneeze, 'pitchoo' or 'pitchaa-tchaa-tchaa'; monotone song, long repetitive phrases of 'tyip-tyip-tyip-tyip'.

TYPICAL FEATURES
It is very likely that the Marsh Tit can only be distinguished from the Willow Tit in the wild by its voice. Willow Tits have a very nasal and drawn out call, a bit like a 'si-si-tchaa-tchaa-tchaa'.

DISTRIBUTION: In deciduous and mixed woodland with dense undergrowth, also in parks and orchards; despite its name it is rarely found in marshes.

FOOD: Small insects and their larvae, spiders; in the winter period eats small seeds, mainly thistle seeds.

NESTING: Nest made mostly of moss in tree holes or nesting boxes; normally only one clutch of seven–nine red-spotted eggs per year.

SIMILAR SPECIES: The extremely similar Willow Tit *(Parus montanus)* ④ is less common and prefers damp woodland.

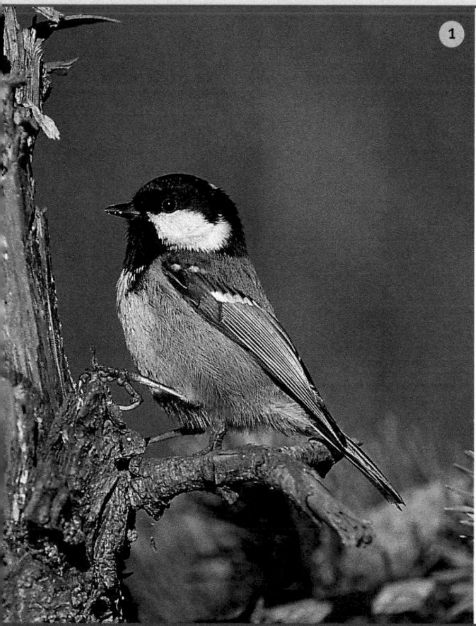
1

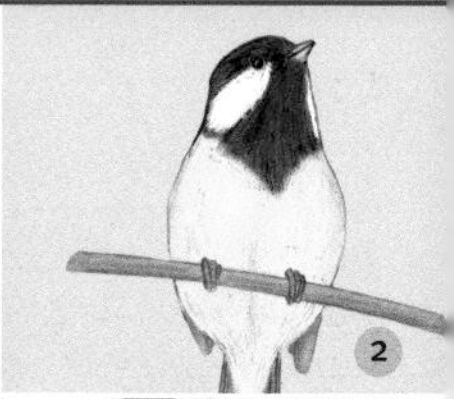
2

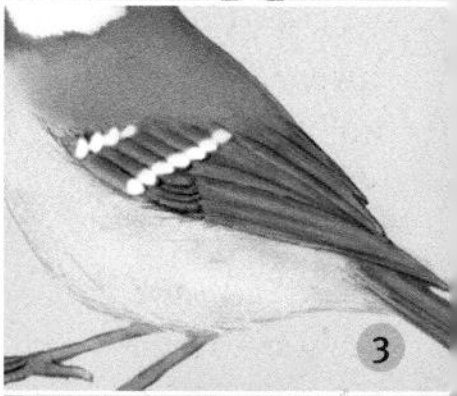
3

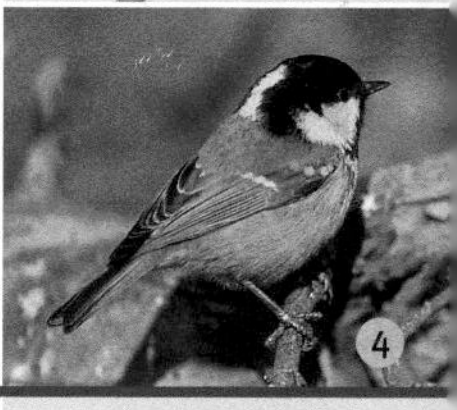
4

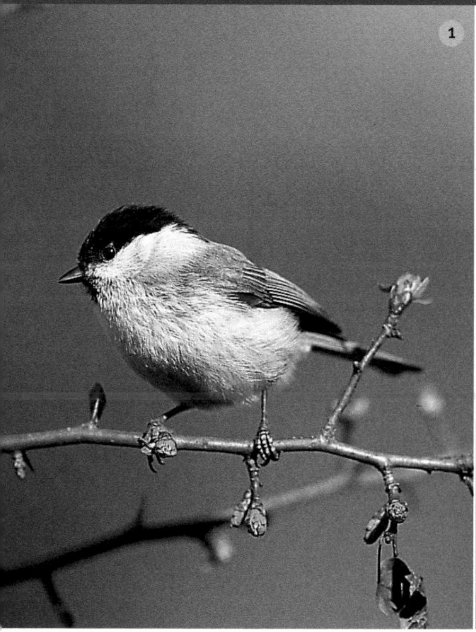
1

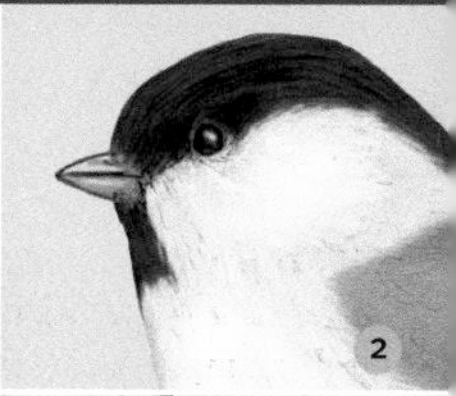
2

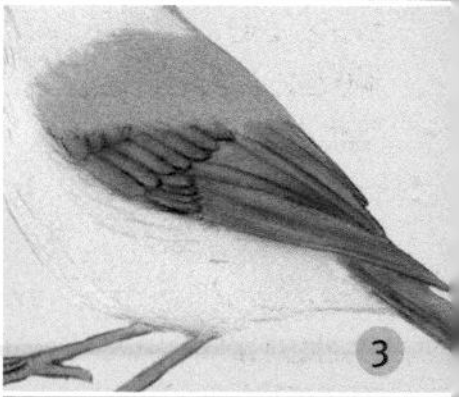
3

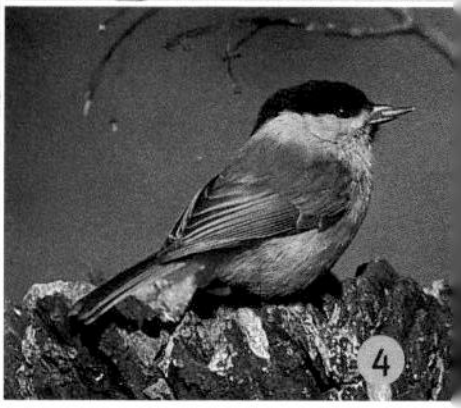
4

Crested Tit
Parus cristatus

DESCRIPTION: The black and white head patterning and noticeable spiky crest ① make this 12cm-long tit very recognisable. Apart from the head, its back is uniform brown ②.

VOICE: Churring call is a bit like a 'urrrrr-r' or 'tsi-tsi-gurrrrr'; song is a changeable series of cooing and trilling sounds like 'zi'i'i-prulull-zi'i'i-prululull', quiet and seldom heard.

DISTRIBUTION: In conifer forests, from lowlands to the treeline in mountainous areas, less frequently also found in parks and gardens with conifers.

FOOD: Mainly insects and their larvae, spiders and other small animals, but also small seeds.

NESTING: Nests in tree holes ④, quite often in woodpecker holes and sometimes also in nesting boxes; five–eight whitish eggs with fine red speckles.

TYPICAL FEATURES

In the juvenile, the characteristic crest is still quite short ③. The typical black and white head markings and the black band around the neck are however noticeable from an early age.

Long-tailed Tit
Aegithalos caudatus

DESCRIPTION: A 12cm-long, chubby looking tit with a noticeably long tail and black-white-pink plumage ①; juvenile has dark brown eye patches and a short tail ③.

VOICE: Three-syllable 'tsi-si-si' call with a piercing pitch; song is a fine, lingering trill and twitter.

DISTRIBUTION: Widespread lowland breeding bird, but it is found nowhere in large numbers.

FOOD: Small insects and spiders.

NESTING: Large, skillfully-constructed, ball-shaped nest ④ of moss and lichen, plant and animal fibres, and quite frequently interspersed with paper and plastic rubbish; eight to 10 whitish eggs.

TYPICAL FEATURES

Central European Long-tailed Tits have a black and white striped head ①. Birds with a plain white head ② belong to a north-eastern sub-species and can sometimes be seen in northern Germany.

OTHER: Outside the breeding season, this sociable bird travels in small flocks. During this period, they often sleep very closely packed, in order to conserve heat.

55

Nuthatch

Sitta europaea

DESCRIPTION: About 14cm long, stocky with short tail; back is uniform grey-blue ④, head has prominent, black eyebrow-stripe ②, in West and Central European birds the underside is orange-brown ①; in North European birds, the breast and belly are white ③.

VOICE: Call is a resounding 'twit twit twit' or 'tsirrr'; sound is loud and penetrating, consisting of a flat flute-like phrases or trilling, like 'tuit-tuit-tuit' or 'fififi'.

TYPICAL FEATURES

The Nuthatch is the only bird in Central Europe that can climb head-first down trees ④.

DISTRIBUTION: Widespread and common breeding bird in Central Europe, in lowlands and mountains; in deciduous and mixed woodland, preferably with oaks; also in parks and gardens.

FOOD: Insects and other small animals, extracted from cracks in bark; also seeds and nuts.

NESTING: Nest of leaves and pieces of bark, constructed in woodpecker holes and nesting boxes; six–eight white eggs with red-brown speckles.

Short-toed Treecreeper

Certhia brachydactyla

DESCRIPTION: Slim songbird, just over 12 cm long, with a relatively long tail; back is bark-coloured ① with a rust-brown rump ②, underside is whitish; the thin, dark beak curves slightly downward ③.

VOICE: High, startlingly loud 'tsee' call; utters an equally loud 'tut tut tut' when alarmed; song is an ascending sequence of thin whistles.

DISTRIBUTION: In deciduous and mixed woodland, in parks and gardens with established trees.

TYPICAL FEATURES

When searching for food, the Short-toed Treecreeper climbs up tree trunks backwards and in a spiral path. Once it reaches the top, it flies down to the foot of the next tree, and begins the pattern again.

FOOD: Insects and their larvae, extracted from cracks in bark.

NESTING: In crevices in tree bark or in nesting boxes, lined with animal hairs and feathers; one–two clutches of four–seven white, red-brown speckled eggs per year.

SIMILAR SPECIES: The Common Treecreeper (*Certhia familiaris*) ④ mainly inhabits extensive conifer forests in high and low mountain ranges.

Tree Sparrow
Passer montanus

DESCRIPTION: Brown with a thick beak ①; at 14cm it is slightly smaller than the House Sparrow. The main difference between the two species is the Tree Sparrow's chocolate-brown crown and a black fleck on the cheek ③; located in the middle of the light patches on the sides of the head.

VOICE: The call of the Tree Sparrow is either a 'tsui-tek-tek-tek', or a 'tschik-tschik-tschok'. Song is a rhythmical, typical sparrow-like chirping.

TYPICAL FEATURES
A thin, white neck band ④ allows the Tree Sparrow to be distinguished from the House Sparrow at a distance.

DISTRIBUTION: Follows humans wherever there are housing settlements, villages, gardens, orchards and parks, still quite uncommon in cities.

FOOD: Seeds, buds, fruits, also insects.

NESTING: Ball-shaped nest of stalks, twigs and feathers, located in tree holes ②, holes in walls or nesting boxes; two–three clutches per year of four–six beige eggs with dense, dark patterning.

House Sparrow
Passer domesticus

DESCRIPTION: About 15 cm long; male has a grey cap on the top of his head, and a black patch on the throat ②; prominent, white wing-bar ① and grey rump ③; female ④ and juvenile have a plain grey-brown colouring.

VOICE: Frequently calls 'tschet-tschet', and its alarm call is a 'tetetete'; song is an even chirping, that can become a clamouring.

TYPICAL FEATURES
The famous rhythmical chirping is the territorial song of the male and is mainly sung from near the nest, from a rooftop or from bushes.

DISTRIBUTION: Found almost everywhere in Europe, where people live. This sociable bird is normally found in groups.

FOOD: Wild plant seeds, insects and their larvae, fruits, berries, cereals and leftovers.

NESTING: Loosely constructed, ball-shaped nest of twigs, stalks, paper and other human rubbish, normally in holes in walls, under roof eaves, in climbing plants, nesting boxes or any sheltered locations; two–three clutches per year of four–six beige coloured, dark patterned eggs.

59

Chaffinch
Fringilla coelebs

DESCRIPTION: Songbird, about 15cm long; both male and female have dual, white wing-bars ②, olive-green rump and white tail edges ③; female ④ is otherwise inconspicuously coloured; male ① is brightly coloured in spring, with a grey-blue cap on the head and a red breast; in winter has duller plumage.

TYPICAL FEATURES
When searching for food on the ground, Chaffinches move backwards.

VOICE: When disturbed, call is a short 'pink'; in flight a short 'yub'; song is warbling and consists of a descending series of notes, repeated many times.

DISTRIBUTION: In all types of woodland, field coppices, hedgerows, parks and gardens, also in city centres; male remains in Central Europe year round; female winters in West or Southern Europe.

FOOD: Seeds and fruits of ground plants and trees, also berries and cereals; in the breeding period also eats insects and spiders.

NESTING: Skillfully constructed, bowl-shaped nest, normally in a tree; two clutches per year of three–six light blue, red-brown speckled eggs.

Brambling
Fringilla montifringilla

DESCRIPTION: Same size and shape as a Chaffinch, but orange-brown on the breast and shoulders; female ④ has a grey-brown head all year; head and back of the male has brown-grey flecks in winter ③, but during the breeding season the head is glossy black ①.

TYPICAL FEATURES
In contrast to the Chaffinch, Bramblings have a bright white rump, particularly visible during takeoff ②.

VOICE: Characteristic screeching call, like a 'wayk', or a 'chuk chuk' when flying; harsher and shorter than the Chaffinch.

DISTRIBUTION: Northern European breeding bird; present in Central Europe in passage or as a winter visitor; found in beech forests, in fields and tips, frequently also in parks and gardens; commonly seen in flocks with other finches and with buntings.

FOOD: Seeds, in particular beech nuts; in spring also buds; in summer insects and other small ground creatures.

NESTING: Nest of moss, stalks, lichen and feathers, normally high in a birch tree or a conifer; five–seven light blue eggs with reddish speckles.

61

Serin | European Serin
Serinus serinus

DESCRIPTION: Only 11 cm long; male ① has a distinctive canary-yellow breast and head, female ③ predominantly grey-brown; both male and female have a bright yellow rump ②.

VOICE: High, trilling 'girlitt' call when in flight; song is light and rattling, performed during flight or from a high perch.

DISTRIBUTION: In light deciduous and mixed forests and in parks and gardens, cemeteries and orchards; widely distributed, but not common; occasionally winters in Central Europe, but normally migrates to southern Europe.

FOOD: Small seeds, also insects.

TYPICAL FEATURES

In spring, the male Serin performs eye-catching fluttering song flights. Its broad wing beats are very similar to those of a bat.

NESTING: Nests in bushes or small conifers; normally two clutches of three–five eggs per year, green or blue-coloured with reddish speckling.

SIMILAR SPECIES: The Citril Finch *(Serinus citrinella)* ④ has less flecking and a grey nape, and is found in light mountain woodland.

Goldfinch
Carduelis carduelis

DESCRIPTION: About 14cm long; both male and female are red-brown with black and white parts ①, a yellow wing-stripe and a red face mask ②; juvenile has a light brown head until its first autumn ④.

VOICE: Mostly a light 'stigelitt' or 'didlitt'; alarm call is a twittering 'faii', and it has a 'tschrrr' call when in conflict with birds of the same species; song consists of cheerful, twittering phrases, interspersed with individual calls.

TYPICAL FEATURES

A Goldfinch can often be identified in flight by its broad yellow stripes on black wings ③.

DISTRIBUTION: On the edges of woods, in hedgerows, orchards, parks and gardens; in autumn and winter can often be seen in groups, looking for seeds left on bushes and shrubs.

FOOD: Mainly tree and shrub seeds, in spring also buds; chicks fed on small insects.

NESTING: Thick-walled, bowl-shaped nest, normally constructed high up in trees and bushes; often two clutches of four–six red-patterned eggs per year.

Greenfinch
Carduelis chloris

DESCRIPTION: About 15cm-long finch, with yellow markings on the wings and tail ②; male ① is a bright yellow-green colour in spring and summer; in autumn and winter it resembles the olive-green female ④; juvenile has flecking on the belly ③.

VOICE: On take-off a ringing 'chew-chew-chew'; alarm-call is a prolonged 'chewee', and in conflict a rasping 'tsrrrr'; song has canary-like trilling, with ringing and whistling sounds.

DISTRIBUTION: In light woods and field coppices and cultivated land with hedgerows;

TYPICAL FEATURES
The Greenfinch female and juvenile can be easily mistaken for a female House Sparrow, but this has no yellow in its plumage.

orchards, and parks and gardens in towns; common almost everywhere.

FOOD: Almost exclusively plant matter: seeds, buds and berries.

NESTING: Nest normally constructed high in bushes or young trees; two–three clutches of four–six eggs per year, with sparse, brown speckles.

Siskin | Eurasian Siskin
Carduelis spinus

DESCRIPTION: 12cm long; male ① has yellow, olive-green and black plumage, dark black head-cap and black bib on the chin ③; female ④ has grey-green back, breast is light with prominent flecking.

VOICE: Flight call is 'di-eh' or 'tiu-li', with stress on the first syllable; song is rapid twittering, which always ends with a prolonged clatter.

DISTRIBUTION: Mainly in mountainous woodland in low mountain ranges up to the

TYPICAL FEATURES
Both male and female birds have a prominent black and yellow wing patterning, clearly visible during flight ②.

treeline. In some places, the Siskin remains in its breeding ground year round, but flocks of Siskins can sometimes be seen migrating from more northerly breeding grounds, and it can inhabit lowlands either in passage, or as a winter visitor.

FOOD: Tree seeds, preferably alders and birches, but also herb seeds; chicks fed on small insects.

NESTING: Nest normally constructed more than 5m above ground on spruce trees; two clutches of four–six light blue, red speckled eggs per year.

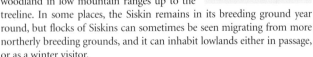

65

Redpoll | Mealy Redpoll
Carduelis flammea

DESCRIPTION: A good 12cm long; male is brown and his breeding plumage has a red-pink coloured breast ①; female ④ is more grey-brown coloured; both have a bright red forehead and a small, black patch on the chin ② as well as dual wing-bars ③.

VOICE: In flight, a rapid and nasal 'tji-tji-tji'; song is twittering and trilling, interspersed with the flight call.

TYPICAL FEATURES
Both the male and female of the Redpoll have a bright red forehead all year round ②.

DISTRIBUTION: In forests near the treeline, recently also in parks and gardens of some towns and cities. A slightly bigger and lighter Scandinavian sub-species can be found in Europe over the winter period in passage or as a winter visitor.

FOOD: Small seeds from deciduous trees and leafy plants; also small insects during the breeding period.

NESTING: Nest normally high in conifers; two clutches of four–six eggs per year, light blue with red and brown flecks and patterning.

Linnet
Carduelis cannabina

DESCRIPTION: About 14cm long; male breeding plumage ① has a red forehead and is easy to identify by its carmine-red breast colouring, divided into two patches ②; its winter plumage is very similar to the plain grey-brown of the female ④; black and white tail is lightly forked ③.

VOICE: In flight, a nasal and somewhat stuttering 'gegegegeg'; song is clear, cheerful and variable, consisting of twittering and trilling, flute-like tones.

TYPICAL FEATURES
The flight of the Linnet is very flat. The bird is normally identified by its continuous and nasal flight call.

DISTRIBUTION: In open countryside with coppices and hedgerows, also in vineyards, orchards, parks and gardens; common in some regions; out of the breeding season, often seen in flocks on fallow and newly harvested fields; also found as a winter visitor from north-eastern breeding grounds.

FOOD: Seeds from leafy plants, less frequently also tree seeds.

NESTING: Nests usually constructed in low hedgerows, bushes or young trees; two clutches of four–six eggs per year; whitish, red speckles.

67

Crested Lark
Galerida cristata

APPEARANCE: About 17cm-long, stocky lark with high, pointed crest, normally held upright ①; brown back, front of throat densely flecked; underside is white ③; tail has no white ④.

VOICE: Soft warbling call, for example 'die-die-dri-e' or 'diu-diu-diur-dli'; whistling and twittering territorial song, consisting of variable, long, frequently repeated phrases.

TYPICAL FEATURES
Unlike the Skylark, the Crested Lark sings from the ground or from a high perch ②.

DISTRIBUTION: Originally an inhabitant of flat, very dry areas, found in Central Europe on dry wasteland, playing fields, industrial sites and road embankments, even in the middle of cities; has become rare in places.

FOOD: Mainly seeds, but also cereal grains, fresh grass tips, small ground insects and spiders.

NESTING: Nests on the ground, often in roadside bushes and embankments, but also on flat roofs; normally two clutches of three–five brown speckled eggs per year.

Skylark
Alauda arvensis

APPEARANCE: Roughly 18cm long, non-descript, brown ground-dwelling bird ① with white belly; feathers on the crown can be raised to form a short, round crest ③.

VOICE: Call sounds like a 'tirrr', 'priutt' or a 'tschrl'; song consists of long, trilling phrases, performed during a lengthy song flight, during which the larch dashes about the sky.

TYPICAL FEATURES
In flight, narrow white banding can be seen on the rear wing edges and on the side tail edges ②.

DISTRIBUTION: Open fields, moors and coastal areas; recently, numbers have decreased drastically in the United Kingdom; only remains year round in Central Europe in places with a mild climate; winters in western or Southern Europe.

FOOD: Insects, larvae, spiders, seeds and plant greenery.

NESTING: Nest made of stalks, hidden in a small ground hollow; normally two clutches of three–four brown speckled eggs per year.

SIMILAR SPECIES: The slightly smaller Woodlark (*Lullula arborea*) ④ is found in only a few scattered locations in Central Europe and, unlike the Skylark, sings from the trees and also at night.

Nightingale
Luscinia megarhynchos

Appearance: A good 16cm long; red-brown back with lighter plumage underneath, without any other patterning ①, rump and tail are a bright red-brown, wings and tail are rounded ②; juvenile has dense flecking ③.

Voice: Alarm call is a whistling 'hweet' or a rough 'karrr'; song is particularly loud and cheerful, and can be heard almost all day and night during the breeding period ④; warbling tune intermingled with an ascending 'hu hu hu hu…'.

TYPICAL FEATURES
The Nightingale hops about on the ground with long jumps. This movement is characterised by regular tail flashes.

Distribution: Once common but now rarer in the United Kingdom; in deciduous woodland, parks, cemeteries and large gardens; winters in Africa.

Food: Insects, spiders, snails, worms; berries in autumn.

Nesting: Nest made from old leaves and stalks, located at or just above ground level, well-hidden in dense undergrowth; four–six olive-green eggs.

Thrush Nightingale
Luscinia luscinia

Appearance: Very similar to the Nightingale, but its back is not red-coloured, but olive-brown ①, breast is beige with cloud-like brownish flecks ③, cheeks enclosed by a pale rim ②; juvenile has heavy flecking.

Voice: When agitated, a deep 'karrr' or a lighter 'hiiied' call; song is slower and deeper than that of the Nightingale, but is also heard throughout the day and night ④.

TYPICAL FEATURES
The Thrush Nightingale can be distinguished from the Nightingale by its lightly patterned breast and brownish flecking or faint striping on its beige-coloured underside.

Distribution: Common in Scandinavia and Eastern Europe but rare visitor to the United Kingdom. Inhabits dense, shady vegetation, preferably near water, but also overgrown gardens and parks; winters in East Africa.

Food: Insects, spiders, snails, worms; berries in autumn.

Nesting: Nest on the ground, constructed from leaves, stalks and twigs, normally in a hollow in dense undergrowth; four–six brownish eggs.

71

Wallcreeper

Tichodroma muraria

APPEARANCE: Blood-red patches on the wings ① and a long, thin beak make this 16cm-long songbird very recognisable. Breeding plumage of the male is a black throat ②, but in the eclipse plumage ③ this is light and identical to the female. With folded wings, the bird is stone-grey and camouflages perfectly among rocks ④.

VOICE: Call is a thin, whistling 'chewee'; song consists of three–five prolonged whistles, descending on the final tone.

DISTRIBUTION: Mainly in the Alps above the treeline, very rarely in the United Kingdom. In winter, in valleys and occasionally in villages and towns.

FOOD: Insects, larvae, spiders, woodlice and millipedes, plucked from cracks in the rocks using its pincer-like beak.

NESTING: Nests in cracks in rocks; four–five whitish, brown speckled eggs.

> **TYPICAL FEATURES**
> When searching for food, the Wallcreeper climbs up rock faces backwards, constantly twitching its wings and displaying its red and white patches.

Northern Wheatear

Oenanthe oenanthe

APPEARANCE: 15–16cm-long ground-dwelling bird; male breeding plumage ① has a black eye-mask and ochre-yellow throat; after moulting in August, plumage has less contrast and is similar to the grey-brown female ④; juvenile has dense flecking ③.

VOICE: When agitated, a tuneless 'tk-tk-tk', often interspersed with occasional 'fiid' sounds. Its chattering, murmuring song is interspersed with light whistling tones. The male usually sings from the ground, but also occasionally during a short song flight.

DISTRIBUTION: Found in mountainous, rocky regions, wasteland and dunes, as well as in valleys; winters in tropical Africa.

FOOD: Insects and spiders, caught on the ground.

NESTING: Nest of grass and moss, in hollows in the ground, rabbit burrows, piles of stones or holes in walls; five–six whitish eggs.

> **TYPICAL FEATURES**
> The upper part of the tail is bright white and has a broad black patterning in the form of an inverted 'T' ②.

73

Waxwing
Bombycilla garrulus

APPEARANCE: About 18cm-long, stocky song-bird with rust-brown to beige-grey plumage; head has a black face-mask and a crest that can be raised and lowered ①, tail has bright yellow tip ③, wings are a contrasting white-yellow-black, with bright red extensions to some secondary flight feathers ②.

TYPICAL FEATURES
The Waxwing's direct flight, with regular gliding periods, is similar to that of the Starling.

VOICE: High, bell-like 'sirrrr' call.

DISTRIBUTION: Breeding bird in northern Scandinavia and Russia, where it inhabits spruce and birch forests; found in the United Kingdom only as a winter visitor, but irregularly and in differing numbers; seen in groups in coppices, parks and gardens containing berry bushes.

FOOD: In the breeding period, mainly insects; in autumn and winter almost exclusively berries ④ and fruit remaining on the trees.

NESTING: Nest of twigs and lichen; three–five white, dark speckled eggs.

Golden Oriole
Oriolus oriolus

APPEARANCE: About 24cm-long, slim song-bird; male ① is sun-yellow with black wings; female and juvenile have a yellow-green back ④, but a whitish underside with dark flecking ③; young males turn yellow in their third year.

TYPICAL FEATURES
The Golden Oriole has an undulating flight a bit like a woodpecker. Its yellow patterning makes it easy to recognise ②.

VOICE: Call of both male and female is a harsh 'krrraee', similar to a Jay, or a 'gewaeh'. The male has a melodic warbling song in spring, which sounds like a 'weela-weeoo'.

DISTRIBUTION: In deciduous and meadow woodland, but also in parks with tall, estab-lished trees, and in large orchards; always perches in the tops of trees. A rare bird of passage in the United Kingdom, though 42 breeding pairs have been recorded. Migrates to tropical Africa in the winter.

FOOD: Large insects, caterpillars, also berries and other fruits.

NESTING: Its skillfully constructed nest is a free-hanging bowl, attached to a fork in a branch. three–five whitish to pink-coloured eggs with sparse dark brown speckling.

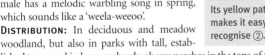

74

Common Quail
Coturnix coturnix

APPEARANCE: Only about 18cm long; camouflages well on the ground due to its earthy-brown, heavily speckled plumage ①; male has ochre-brown breast and black throat ②.

VOICE: Three-syllable, rhythmical, prolonged territorial call, for example 'pick-wer-wick', normally at dusk.

DISTRIBUTION: In open, cultivated countryside with meadows and fields; some winter in the Mediterranean, others in tropical Africa.

TYPICAL FEATURES
The Common Quail almost always runs away on foot. It is only rarely seen flying, just above ground level with rapid wing beats.

FOOD: Insects, snails, worms, shoots and seeds of leafy vegetation and cereals.

NESTING: One–two clutches per year of six to 12 yellowish, dark speckled eggs, laid in a nest on the ground. Young are nidifugous.

SIMILAR SPECIES: The slightly larger Corncrake *(Crex crex)* ④ is sometimes found in the same habitat as the Common Quail. Its flight is undulating, with weak wing beats and hanging legs ③.

Pied Wagtail | White Wagtail
Motacilla alba

APPEARANCE: About 18cm long, slim songbird with long legs; male has contrasting black-white-grey plumage in summer ①, in winter is slightly duller like the female ③; long tail with white edges ②; juvenile plumage brown-grey ④.

VOICE: Call sounds like a 'pewitt', 'ziwlitt' or 'zitt'; song is a cheerful, rapid twittering, but it is only rarely heard.

TYPICAL FEATURES
A rapid and cheerful song, accompanied by constant head-bobbing and tail-flashing, is typical of the Pied Wagtail.

DISTRIBUTION: Prefers to live near water, but also found in isolated farms, in villages and towns, parks and gardens; common in many places. A small number winter in Central Europe, but most migrate to the Mediterranean or even further south, as far as central Africa.

FOOD: Insects and their larvae, spiders.

NESTING: Nest is in shallow hollows near water, but also often in buildings, sheds or wood stores; normally two clutches per year of five–six light green, dark speckled eggs.

77

Yellow Wagtail
Motacilla flava

APPEARANCE: 17cm-long songbird with long tail and legs; in summer the breast and belly of the male is bright yellow, its back is olive-brown, and its head can be grey with a white eyebrow-stripe ① or yellow-green ② depending on the sub-species; the winter plumage and the female bird ④ have duller colours and less contrast; juvenile plumage ③ has an olive-brown back and dirty-white belly.

VOICE: Call is a 'psieh' or 'psuip'; song consists of a series of syllables similar to the call.

TYPICAL FEATURES
Yellow Wagtails often sit on fence posts or low bushes. Like all wagtails, they are noticeable for their constant tail-flashes.

DISTRIBUTION: Damp meadows with short grass, fields, often near villages; common in lowlands, but rare in high hill country; winters in Africa.

FOOD: Insects and their larvae, caught on the ground.

NESTING: Flimsy nest, hidden in ground vegetation; one–two clutches per year of four–six eggs, densely speckled with reddish-brown.

Grey Wagtail
Motacilla cinerea

APPEARANCE: 18–19cm-long, slim wagtail with grey back; male underside ④ is far more bright yellow than that of the female ①; throat white ②, and is only black in the male breeding plumage; in flight its white tail-edges and white wing-stripes are visible ③; juvenile has a cream-coloured breast.

VOICE: Call is a loud and sharp 'tsick-kick' or 'tsiss-tsiss'; alarm-call a shrill 'sisiht'; song is a series of whistling tones, trills and calls.

TYPICAL FEATURES
Unlike the Yellow and Pied Wagtails, the solitary Grey Wagtail is almost never seen in large groups.

DISTRIBUTION: Widespread. Found in mountainous regions, even above the tree line; near clear, quick-flowing rivers and streams, only rarely found far from water. Some birds winter in Britain or in the Mediterranean, others remain in their breeding grounds.

FOOD: Insects, spiders and small worms.

NESTING: Nests in holes in bank vegetation, in cracks in rocks, also under bridges; normally two clutches per year of four–six red-brown speckled eggs.

79

Ringed Plover
Charadrius hiaticula

APPEARANCE: 18–20cm long, back sandy-brown, white belly ①; breeding plumage has a contrasting black and white face and an orange beak with black tip ②; winter plumage is brown and white ④; a white wing-stripe ③ is visible during flight.

VOICE: Soft, two-syllable 'tui-it'; during courtship, call is a prolonged 't'wuieh t'wuieh …' or 'driu driu driu …', sung in flight.

DISTRIBUTION: Common breeding bird on the North Sea and Baltic Sea coasts; winters on the Mediterranean coast, also seen regularly in passage on the banks of rivers and lakes in Central Europe.

TYPICAL FEATURES
Like all plovers, the Ringed Plover walks with very rapid steps. When standing, the Ringed Plover makes jerky movements, followed by a sort of 'bow'.

FOOD: Insects, larvae, small crabs, worms and snails.

NESTING: Nest is a simple scraped-out hollow on the ground; normally two clutches per year of on average four sand-coloured, dark speckled eggs.

Little Ringed Plover
Charadrius dubius

APPEARANCE: Similar to the Ringed Plover, but slightly smaller and slimmer, with a black beak and a thin, citrus-yellow ring around the eyes ②; juvenile ④ has a pale brown head, similar to the adult bird in its winter plumage.

VOICE: When agitated, a high, whistling 'piu'; during courtship it makes bat-like, fluttering song flights, uttering a rapid trill, a bit like a 'grugru gru …' or a 'tria tria tria …'.

TYPICAL FEATURES
The Little Ringed Plover does not have the white wing-stripe ③ of the Ringed Plover.

DISTRIBUTION: Widespread, but not particularly common inhabitant of lowland areas; sand or stony riverbanks and lake shores, gravel and sand flats; migrates to sub-Saharan Africa for the winter.

FOOD: Beetles, small crabs, snails, worms and other small animals found on the ground.

NESTING: Nest is simply a shallow hollow on a sand or gravel bank near water ①; normally four eggs, well-camouflaged by their dark speckling. As with all wading birds, the young are nidifugous.

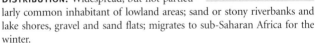

81

Common Sandpiper

Actitis hypoleucos

APPEARANCE: 19–21cm long, back is olive-brown with faint patterning; belly is pure white ①. In flight, a white wing-stripe can be seen on the upper surface of the wing ②.

VOICE: Light, piercing call, a bit like a 'hieh-di-di', with the stress on the first syllable; when surprised, call is a 'iiit'; male has a trilling courtship song which he sings whilst in the air.

DISTRIBUTION: On lightly vegetated, gravelly and sandy banks of rivers and lakes, from lowlands to high mountains; winters in the Mediterranean.

TYPICAL FEATURES

The Common Sandpiper likes to perch on semi-submerged stones or floating wood, constantly waggling its rear.

FOOD: Insects, spiders, small worms and small crabs.

NESTING: Ground nest with little padding, normally hidden under vegetation; three–five ochre-brown to reddish eggs with dark brown speckles.

SIMILAR SPECIES: The slightly bigger Green Sandpiper (*Tringa ochropus*) ④ has a prominent white rump and banded tail ③.

Black Tern

Chlidonias niger

APPEARANCE: 22–24cm-long, grey seabird; during the breeding season, the front part of the body is almost black, with a soot-black head ①; tail white and only slightly forked ②; juvenile plumage has white head patterning ③, similar to the eclipse plumage of the adult ④.

VOICE: Short, scratchy call, a bit like a 'kriairr', 'kirr' or 'kraek'.

DISTRIBUTION: Nests in colonies in marshes and heavily silted lakes with lush vegetation; more common in Northern Europe; winters in tropical Africa; can be seen in passage on all types of inland waterways and sea coasts.

TYPICAL FEATURES

Black Terns fly just above the water when looking for food and often remain hovering in the air. They normally pluck their food from the surface of the water.

FOOD: Water insects, their larvae and other small water creatures.

NESTING: Nest constructed on broken reeds, floating leaves or grass stems protruding from the water; two–three clay-coloured or olive-brown eggs with large, dark speckles; both partners incubate the eggs and look after the young chicks.

Wood Sandpiper
Tringa glareola

APPEARANCE: 19–21cm-long, graceful wading bird; back has fine, pearl-like flecking, belly is white ①; legs pale green.

VOICE: A piercing 'giff giff giff' call during take-off; courtship song consists of long, soft 'dlie dlie dlie' phrases.

DISTRIBUTION: Nests in moors and marshes in Northern Europe, but quite rare in Central Europe; seen quite frequently in passage, both on the coast and inland; winters in central and southern Africa.

FOOD: Insects and their larvae, worms, snails, crabs.

TYPICAL FEATURES

In flight, the Wood Sandpiper displays a square, white rump ②. The rump of the Greenshank is more elongated and wedge-shaped ③.

NESTING: Nest usually on the ground between dwarf bushes; normally four eggs; egg colour is pale green with irregular dark speckles.

SIMILAR SPECIES: The Greenshank *(Tringa nebularia)* ④ is very similar to the Wood Sandpiper in terms of colouring, but is noticeably larger.

Dunlin
Calidris alpina

APPEARANCE: 16–22cm-long, very sociable wading bird; back is red-brown in summer ①, but in winter is more grey ④; legs are black, as is the relatively powerful and slightly down-curved beak ②.

VOICE: In flight, a forced, rolling 'triurrrr'; a trilling song during its courtship flight; on the nest a quiet 'fififi' or 'feutt'.

TYPICAL FEATURES

Both male and female Dunlins have a notice-able black patch on the belly during the summer ③.

DISTRIBUTION: Breeding grounds in the Arctic tundra and in northern mountain regions, but also sandflats and marshes on northern coasts; occasionally also found on the North and Baltic Sea coasts; during the migration period (March to May, and July to early November) is often found in giant flocks on mud-flats. Some birds winter in Central Europe, but most migrate to West European coasts and to the Mediterranean.

FOOD: Insects, mud worms, small mussels and snails.

NESTING: Nest hidden in dense ground vegetation; three–five beige-coloured, brown speckled eggs. Young are nidifugous.

85

Common Kingfisher
Alcedo atthis

APPEARANCE: A 16–17cm-long, stocky bird; underside is orange-red ④, back is blue and turquoise-coloured with a metallic shine ①; the lower beak of the female is a reddish colour ③.

VOICE: Sharp, piercing 'tiiiit', increasing to a 'titititi' when agitated.

DISTRIBUTION: Beside rivers and streams with clear water and steep banks with dense vegetation; sometimes also by ponds and lakes; solitary; not common in any location.

FOOD: Small fish, caught during plunging dives; also water insects and their larvae, tadpoles, small crabs.

NESTING: Nest is in a self-dug, horizontal burrow in steep banks; burrows can be up to 1m long; normally two, but sometimes three clutches per year of on average six–seven pure white eggs.

TYPICAL FEATURES
The Kingfisher will often sit for hours on branches or twigs, hanging over water, spying for its prey. It also has a characteristic, arrow straight flight, normally just above the surface of the water ②.

White-throated Dipper
European Dipper
Cinclus cinclus

APPEARANCE: About 18cm-long, stocky songbird with short tail, often held stiffly upright ②; chocolate-brown plumage with bright white patch on the throat and breast ①. Juvenile has grey-brown, densely flecked plumage that looks slightly like scales, and a lighter, but not clearly demarcated breast patch ③.

VOICE: Sharp 'tsitt' or rough 'schraett-schraett'; song is a rough twittering and sharp chattering, performed by both male and female; also heard in winter.

TYPICAL FEATURES
The White-throated Dipper often stands on stones in the middle of the water, making dipping movements.

DISTRIBUTION: Widespread in the Highlands and low mountain ranges, absent from lowland areas; lives by quick-flowing, clear streams and rivers to an altitude of 2,000m.

FOOD: Water insects and their larvae, small crabs, occasionally small fish; caught while swimming or even diving.

NESTING: Large, ball-shaped nest of moss, with side entrance ④, often used over several years; two clutches per year of four–six white eggs.

87

Red-backed Shrike
Lanius collurio

APPEARANCE: About 17cm long, with black, hooked beak ②; male has black eye-stripes and bright red-brown back ①; the underside is white, but with slight pink colouring; head and rump are light grey, tail black and white ③; female ④ and juvenile are mainly brown with a paler belly.

VOICE: Staccato 'dschae' or a tuneless 'trrt-trrt'; when the nest is disturbed, a harsh 'tek'; song is a light chattering.

DISTRIBUTION: In open countryside with lots of hedgerows, meadows, the edges of woods and moor and heathland, preferably with dense thorny bushes; migrates to Eastern and Southern Africa for the winter.

FOOD: Large insects, small frogs, lizards, mice, young birds.

NESTING: Bowl-shaped nest, normally low down in dense thorn bushes; four–six eggs, with a crown of brown speckles at the blunt end.

> **TYPICAL FEATURES**
> The Red-backed Shrike often impales its prey on thorns, making it easier for the bird to shred its food into more manageable-sized pieces.

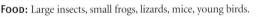

Great Grey Shrike
Northern Shrike
Lanius excubitor

APPEARANCE: About 24cm-long, black and white songbird with a silver-grey back ②, black eye-mask and a hooked beak ①; juvenile is grey-brown with pale tidemark patterning on the underside ③.

VOICE: When the nest is disturbed, a harsh 'faed faed' call; song consists of long, whistling sounds and harsh, chattering tones.

DISTRIBUTION: Open countryside with clumps of trees and bushes; established orchards; widespread, but not common. Many winter in Europe, and are joined by other birds from north-eastern breeding grounds.

FOOD: Large insects, mice, small birds up to the size of a bunting, lizards; large prey is impaled on thorns, branch tips, or wedged into a branch fork, in order to make it easier to shred ④.

NESTING: Nest is normally high in a lone tree or bush; five–six whitish eggs, densely covered with large, brown speckles.

> **TYPICAL FEATURES**
> The great grey Shrike looks for prey from a high perch or while hovering. He then dives in for the catch like a bird of prey.

89

Great Spotted Woodpecker
Dendrocopos major

APPEARANCE: 22–23cm-long woodpecker with black-white-red plumage; characteristic long white shoulder-patches ② and a bright red rump; male ① has a scarlet stripe on the back of the neck; female ④ has no red on the head; both male and female juveniles have a red crown ③.

VOICE: Metallic sounding 'kick' call, repeated in rapid succession when alarmed; in spring, the male will fly after the female as part of a courtship flight, with a husky 'raeraerae' call.

TYPICAL FEATURES
The drum roll used by both males and females to mark their territory is short, and always stressed at the beginning.

DISTRIBUTION: In all types of woodland, also often in parks and gardens with tall trees, even in city centres; common.

FOOD: Wood-dwelling insects and their larvae, also tree sap and the eggs and chicks of other birds; in winter eats mainly tree seeds.

NESTING: The breeding pair work together to peck out a new nesting hole each year; normally five–seven white eggs; young raised by both parents.

Middle Spotted Woodpecker
Dendrocopos medius

APPEARANCE: 20–22cm long, slightly smaller than the Great Spotted Woodpecker; back is a contrasting black and white; white underside has fine flecking ①; rump is soft pink ②.

VOICE: Individual, very faint 'kuck' call, occasionally an accelerating series of calls; territorial song consists of nasal, screeching sounds; drums only very occasionally.

TYPICAL FEATURES
Unlike the Great Spotted Woodpecker, the Middle Spotted Woodpecker doesn't have a red neck, but a completely red crown.

DISTRIBUTION: Deciduous and mixed woodland, preferably with oaks and hornbeams; parks and large gardens with suitable tree stock; only found in lowlands and not in mountainous areas.

FOOD: Insects, plucked from leaves and twigs, and also more rarely pecked out of wood; in winter also eats tree seeds.

NESTING: Nesting holes in hollow tree trunks; five–six white eggs.

SIMILAR SPECIES: The Lesser Spotted Woodpecker (*Dendrocopos minor*) is only 15cm long and only the male ④ has a red cap. The female ③ has no red in the plumage, but has an ochre-yellow crown.

Three-toed Woodpecker

Picoides tridactylus

APPEARANCE: The black and white plumage of this 22cm-long woodpecker has intense patterning. A broad white stripe extends from the neck to the rump on the back of the bird ③. The male ① has yellow markings on the crown, but these do not appear on the female ②. Easily identified by their feet, which have just three toes, rather than the usual four.

VOICE: Soft, muffled 'ugg' call, only rarely heard; extended, rattling drumming can be heard frequently.

TYPICAL FEATURES

In order to reach the tree sap that it loves to feed on, the Three-toed Woodpecker drills a row of holes in the bark of conifers ④.

DISTRIBUTION: Found in the Alps, in Alpine forests and in dense woodland such as the Black Forest, but not found in Britain; prefers spruce forests at high altitudes of between 800 and 1,700m.

FOOD: Insects and larvae living in wood and under the bark; in early summer also eats tree sap.

NESTING: Pecks a new tree hole each year; three–four white eggs.

Wryneck

Jynx torquilla

APPEARANCE: The Wryneck is 16–17cm long and actually belongs to the woodpecker family, despite the fact that its relatively short beak ② and brown colouring make it look more like a songbird. Plumage has a bark-like patterning ①, and a black-brown stripe running the length of the back ③.

VOICE: Loud, somewhat melancholy-sounding 'gjaegjaegjae' call, rising at the end; sometimes also a hissing 'gschrih'.

TYPICAL FEATURES

During courtship, but also when threatened, the Wryneck bends its head backwards, turning it from side to side and hissing.

DISTRIBUTION: Relatively widespread in Central Europe, but rarely seen; solitary, inhabiting light deciduous woodland, field coppices, avenues, orchards, cemeteries, parks and gardens with established deciduous trees; winters in tropical West and Central Africa.

FOOD: Primarily ant pupae, collected on the ground; occasionally also other small insects and their larvae.

NESTING: Six to 10 white eggs are laid in tree holes, without any nest lining ④ (old, abandoned woodpecker holes); also lays in nesting boxes.

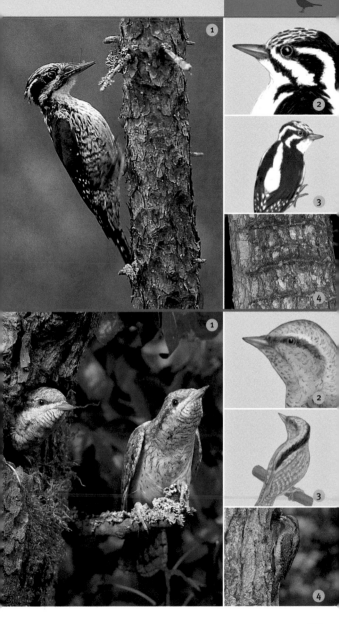

93

Redwing
Turdus iliacus

APPEARANCE: Roughly 21cm long, with rust-brown flanks and a prominent, yellow-white eyebrow-stripe ①; black flecking on its light breast, roughly arranged in lines along the belly ④; in flight, the red-brown colour beneath the wings can be seen ③.

VOICE: Flight call is a high, prolonged, somewhat croaky 'tseep'; song is a descending series of flute-like tones, with a short twittering.

DISTRIBUTION: Breeding bird in northern taiga woodland; found in Central Europe mainly in passage or as a winter visitor; seen searching for food in meadowland, also in gardens and parks with berry bushes.

FOOD: In autumn and winter eats mainly berries ②, but also small animals, for example worms or snails.

NESTING: Sturdy nest constructed in lower branches of trees and bushes; also sometimes on the ground; five–six eggs with fine, red-brown pattern.

> **TYPICAL FEATURES**
> During still nights in March, April and October, you can hear the characteristic flight call of the migrating Redwing.

Song Thrush
Turdus philomelos

APPEARANCE: About 23cm-long thrush; back is dark brown, underside creamy-white with numerous dark flecks ①; face has no noticeable eyebrow-stripe ③.

VOICE: In flight, often a short, sharp 'tsipp'; in danger, a piercing chatter; song is resounding, consisting of various phrases; song can be heard morning and evening, performed from the top of tall trees or other high perches.

DISTRIBUTION: In woods with heavy undergrowth and in field coppices, also common in gardens and parks; in mountain regions to an altitude of 2,000m; found year round in Britain though some birds migrate to the Mediterranean.

FOOD: In spring mainly earth worms, in summer caterpillars, in autumn eats berries and fruit; eats snails year round ④.

NESTING: Sturdy, bowl-shaped nest, often in young spruce trees, close to the trunk; two clutches per year of four–six turquoise-coloured eggs (p23).

> **TYPICAL FEATURES**
> In flight, the Song Thrush displays a rust-yellow coloured patch on the underside of the wing ②. Unlike the Mistle Thrush (p122) its flight is relatively straight.

95

Blackbird
Turdus merula

APPEARANCE: About 25cm long; male has soot-black plumage, yellow beak and yellow ring around the eye ①, female is a uniform dark brown ②; juvenile is a light brown and has dense flecking ③; partial albinos with a greater or lesser number of white patches are often seen in towns ④.

VOICE: Call is a sharp 'tjock' or 'tic tic', or a piercing chattering when agitated; song is very melodic and variable.

DISTRIBUTION: Mainly in parks, gardens and city centres, the commonest bird in Britain.

FOOD: Earthworms, snails and insects; in autumn also eats berries and fruit.

TYPICAL FEATURES
The loud and melodic song of the Blackbird can be heard as early as late winter. It is performed at dawn from a high perch, e.g. tree tops, roofs or television antennae.

NESTING: Nest strengthened with clay, in bushes and hedgerows; also on the outside of buildings, e.g. in trellises or on balconies; two–three clutches per year of three–five blue-green eggs, with fine, brown speckling (p23).

European Starling
Sturnus vulgaris

APPEARANCE: About 21cm long; plumage black with fine white flecks; during breeding season, male plumage has a violet and green metallic sheen ①; after moulting in autumn, plumage is particularly densely covered in pearl-like flecks ④; juvenile is brown ②.

VOICE: Hoarse 'rraeh' call; when in danger, a sharp 'pett pett'; song is an incessant chattering, interspersed with whistling and clicking sounds. When singing, the Starling flaps its wings energetically ③.

TYPICAL FEATURES
The Starling has a quick, waggling walk when on the ground. This is in contrast to the Blackbird, which hops on two legs.

DISTRIBUTION: In deciduous and mixed woodland, field coppices, parks and gardens; common everywhere apart from in mountain regions; outside the breeding season, often seen in large flocks; winters mainly in Southern Europe, but some remain in Britain.

FOOD: Insects, worms, snails, also berries and fruit.

NESTING: Nests in holes in trees, walls and cliffs, also in nesting boxes; one–two clutches per year of four–seven turquoise-green to light blue eggs.

97

Reed Bunting
Emberiza schoeniclus

APPEARANCE: 15cm long; male breeding plumage has a black head, with a contrasting white band around the throat and a white beard-like stripe ①; after the autumn moult, male resembles the female ②, with a brown head and a dark beard-like stripe on a white throat ③.

TYPICAL FEATURES
Reed Buntings regularly twitch their tails, which have distinctive white edges ④.

VOICE: Most common call is a sharp, descending 'tsieh', and a trilling call; territorial song consists of short, stammering phrases, a bit like a 'tsie tsie tsie tui tsirri'.

DISTRIBUTION: In shingle and muddy areas, such as land surrounding ponds and lakes, on river banks or in marshes; short-distance migrant, but some winter in Britain in places with a very mild climate.

FOOD: Seeds, mainly grass seeds, but also insects.

NESTING: Nests mostly on the ground, hidden in tall tufts of grass; two clutches per year of five–six pale lilac-coloured or brown eggs, with dense violet or black patches and patterning.

Yellowhammer
Emberiza citrinella

APPEARANCE: Songbird, 15cm long, with a relatively long tail; especially during the breeding season, the male has a bright yellow head and breast ①; on the winter plumage ③ this yellow colour is disguised by olive-green feather edges; female is a paler yellow, with dense flecking on the head and breast ④.

TYPICAL FEATURES
When the Yellowhammer takes off, its rust-brown rump and its bright white tail edges become visible ②.

VOICE: Call is a 'tsrik', 'tzue' or 'tsurr'; song consists of a series of unchanging high tones with a prolonged final note, that has been described as 'little-bit-of-bread-and-no-cheese'.

DISTRIBUTION: In small meadows with hedgerows and bushes, also in narrow woodland strips, and in parks and gardens on on the edge of towns.

FOOD: Seeds, buds and other plant greenery; in summer eats many insects.

NESTING: Bowl-shaped nest of stalks, leaves and moss, near the ground and hidden in bushes; two clutches per year of three–five delicate grey or dark red speckled eggs.

99

Bullfinch

Pyrrhula pyrrhula

APPEARANCE: At 16cm it is scarcely larger than a House Sparrow, but its plump build makes it look significantly larger; male has a bright red underside, black, white and grey back and a black head-cap ①; female has red-brown underside ④; juvenile has no head-cap ③.

VOICE: Soft and melancholy 'tjui' or 'iup' call. The song is a quiet, inconspicuous twittering chatter, mixed with screeches and warbling tones.

TYPICAL FEATURES
In flight, the main characteristic feature of the Bullfinch is its bright, white rump ②.

DISTRIBUTION: Widespread and common from lowlands to mountain ranges; in conifer and mixed woodland, field coppices and hedgerows, parks and gardens.

FOOD: Buds, seeds and berries; during the breeding season, also insects.

NESTING: Loose nest of twigs and moss, hidden in thick bushes and conifers; normally two clutches per year of four–six light blue, dark speckled eggs. Bullfinch pairs mate for life.

Hawfinch

Coccothraustes coccothraustes

APPEARANCE: 18cm-long songbird with cinnamon-brown, black and white plumage and extremely thick beak, blue-grey in summer ①, and bone-coloured in winter; in flight, a broad, white wing-stripe and white rump is visible ②.

VOICE: Short, sharp 'tsicks' or 'tsittit' calls often reveal the presence of this retiring, treetop-dwelling bird. Stammering song consists of call variations and nasal tones; only rarely heard.

TYPICAL FEATURES
The Hawfinch's heavy beak means that it is rarely confused with other species of Central European bird.

DISTRIBUTION: In deciduous and mixed woodland, but also in many parks and gardens with tall, established deciduous trees.

FOOD: Seeds from deciduous trees, mainly hornbeams and maples.

NESTING: Large nest of twigs, stalks and thin roots, constructed high in deciduous trees; four–six brown-grey, dark patterned eggs.

SIMILAR SPECIES: The Common Crossbill (*Loxia curvirostra*) also has a very powerful beak, but with tips that cross noticeably. The male is red ④, the female yellow-green-coloured ③.

101

Nightjar
Caprimulgus europaeus

APPEARANCE: About 25cm long and active at dusk and at night; bark-coloured plumage ①; extremely small beak ④; wings are long, and on the male these are flecked with white ②, on the female they are plain brown ③.

VOICE: During the courtship period, the male performs a fluctuating churring trill for hours on end, from the vantage point of a high perch. Flight call is a short 'coo-ick'.

DISTRIBUTION: Only in dry and warm lowland areas, preferably on the edge of pine woodland with small areas of heathland; winters in Africa.

FOOD: Nocturnal insects, caught in flight.

NESTING: Always two densely speckled eggs; not laid in a nest but on a vegetation-free patch of ground. Chicks remain seated in the hatching area, despite the fact that they are able to walk immediately.

TYPICAL FEATURES
If the Nightjar is disturbed on its perch during the day, it normally flies just a short distance away and settles down to sleep on a new branch.

Common Swift
Apus apus

APPEARANCE: 16cm long, similar to a swallow, with thin, sickle-shaped wings ④, which extend far beyond the forked tail when bird is perched ③; apart from a whitish area on the chin and throat ②, the Swift is entirely black-brown ①; juvenile has light flecking.

VOICE: The high and shrill 'srieh' call of this sociable bird can often be heard.

DISTRIBUTION: Often nests in towns and villages; when looking for food, likes to fly over open countryside; winters in tropical Africa.

FOOD: Small, flying insects.

NESTING: Nests in colonies high up in cliff crevices, or in cracks in walls and under roof eaves; two–three white eggs; nest is constructed and chicks are raised by both parents.

TYPICAL FEATURES
The Common Swift is an adept flyer that flashes through the air with rapid, shallow wing beats. Unlike the Swallow, it never folds its wings during flight.

OTHER: Apart from when they are nesting and raising their young, the Swift is almost constantly airborne – even when asleep.

103

Little Owl

Athene noctua

APPEARANCE: 21–23cm long, stocky, with a short tail; dark brown with a dense white flecked patterning ①; eyes sulphur-yellow; juvenile is paler brown ②.

VOICE: Territorial song is a prolonged 'hoooee' or 'keew'; alarm call is a short 'kiu' or 'quiep'; when the nest is disturbed, utters a loud 'kiff kiffkiff' and snaps its beak loudly.

DISTRIBUTION: Has become rare in many places; mainly found in open, cultivated lowland areas with scattered copses.

FOOD: Mainly mice, but also small birds, frogs, lizards, large insects, spiders, earthworms.

NESTING: Three–five white eggs, laid in a tree hole or a wall crevice.

SIMILAR SPECIES: The Boreal Owl (*Aegolius funereus*) ④ is nocturnal and remains hidden during the day, normally in dense conifers. Its round facial patterning extends high above the eyes ③.

> **TYPICAL FEATURES**
> Although mainly active at dawn and dusk, the Little Owl is also often seen during the day. When agitated, it twitches visibly.

Eurasian Pygmy Owl

Glaucidium passerinum

APPEARANCE: 16–18cm long; Europe's smallest species of owl; round body; relatively small, yellow eyes; back is chocolate-brown with white, flecked patterning ①, underside is white with rows of dark speckles ②.

VOICE: Territorial song is a series of whistling tones, a bit like a 'hiub hiub', or a vibrating 'hiub iu-iu-iu-iu-iu…'; in autumn and spring, mainly heard at dusk, but occasionally during the day; in autumn also has an additional call, which is a series of ascending notes.

DISTRIBUTION: In conifer and mixed woodland in the Alps and in medium height mountain ranges.

FOOD: Mainly mice, but also small birds.

NESTING: Three–seven white eggs; nests in abandoned woodpecker holes ④.

OTHER: The Eurasian Pygmy Owl is active at dawn and dusk and catches other birds up to its own body size; prey caught on the ground or in mid-air. Excess food is often stored in woodpecker holes.

> **TYPICAL FEATURES**
> At dawn and dusk, this tiny owl often perches at the tops of a conifer tree and whistles loudly. When agitated it raises its short tail ③.

105

Golden Plover
Pluvialis apricaria

APPEARANCE: Roughly 28cm-long, sociable wading bird; back is dark brown with golden yellow flecking; face and underside of breeding plumage is normally a uniform black ③, but the throat sometimes has some flecking ①, winter plumage is a dirty white ④.

VOICE: Soft, flute-like 'klew' call, becomes a two-syllable 'klew-ee' when agitated. During courtship the male performs a song flight, accompanied by a 'perr-ee-oo, perr-ee-oo'.

TYPICAL FEATURES

In flight, the Golden Plover displays the light underside of its wings and bright white wing bases ②.

DISTRIBUTION: Common breeding bird in northern regions; small number of breeding pairs in Central Europe; common in spring and autumn when in can often be seen as a passage migrant on the North Sea and Baltic Sea coasts in great flocks; inland normally seen in smaller numbers.

FOOD: Insects and larvae, spiders, snails, also berries.

NESTING: Three–four light brown, dark speckled eggs are laid in a hollow amid ground vegetation. Young are nidifugous.

Lapwing
Vanellus vanellus

APPEARANCE: About 30cm-long, sociable wading bird; from a distance appears to be black and white, but close-up it has a green, metallic shimmer; in flight a cinnamon-coloured rump is visible ②; male breeding plumage has long feathered crest and black throat ①; the winter plumage of male and female birds has white throat ④; female has shorter crest ④; juvenile crest is yet shorter ③.

VOICE: Two-syllable, melancholy sounding call, a bit like a 'pee weet'; during the male display flight, call is a 'weet weet weet kchiuweet'.

TYPICAL FEATURES

Male Lapwings perform spectacular display flights involving a twisting, rolling dive. This display is to mark its breeding territory and attract a female.

DISTRIBUTION: Common in lowland silt-covered areas, moors, meadows with short grass and cultivated areas; migrates south when the weather begins to turn cold; during mild winters remains in Central Europe.

FOOD: Insects, earthworms, snails.

NESTING: Normally lays four eggs; olive-brown with black speckles. Eggs laid in a hollow in the ground. Young are nidifugous.

107

Hoopoe
Upupa epops

APPEARANCE: About 28cm long, with short legs; head has a large feathered crest ④, which is fan-like with black tips when extended ①; black and white striping on the wings, back and tail ②, rest of the body has pale, cinnamon-coloured plumage.

VOICE: The Hoopoe's three-syllable call carries far and sounds like a muffled 'bub-bub-bub'. When disturbed, utters a screeching 'shraeee'.

DISTRIBUTION: In light woodland, parks and open, cultivated land; rare in the United Kingdom; winters in Near East and Africa.

FOOD: Large insects, snails, worms, occasionally frogs or lizards.

NESTING: Nest normally contains five–eight white eggs, usually in a hole in a tree ③, or sometimes in cliff and wall crevices, or in holes in the ground.

TYPICAL FEATURES

The flight of a Hoopoe has irregular wing beats and resembles that of a butterfly. On landing and when agitated, the Hoopoe raises its crest ①.

Cuckoo
Cuculus canorus

APPEARANCE: 32–34cm long, with a long tail; back is grey ①, but occasionally, especially on the female, also red-brown ③; underside has close striping, similar to that of a Sparrowhawk.

VOICE: The territorial song of the males is the famous cuckoo-ing call. During the breeding season, the female makes a loud trilling sound; the begging call of the chicks is a piercing 'srie srie…'.

DISTRIBUTION: Most common in semi-open, varied landscapes, where there are a lot of songbirds, also in large parks; winters in tropical Africa.

FOOD: Mainly caterpillars, even very hairy ones avoided by other birds; also eats large insects.

NESTING: The Cuckoo has parasitic nesting habits, and lays as many as 20 eggs, one at a time, in the nests of various species of songbird. The unwitting host then incubates and raises the Cuckoo chick ④.

TYPICAL FEATURES

The flight profile of a Cuckoo is similar to that of a bird of prey. The silhouette displays long, narrow and pointed wings ②.

Woodcock | Eurasian Woodcock
Scolopax rusticola

APPEARANCE: 32–34cm long and stocky, with a long beak ②; plumage has light and dark brown patterning, looks like fallen leaves ④.

VOICE: During the male courtship flight, a muffled 'og og og', followed by a light, high-pitched 'chee-wick chee-wick'; rarely heard outside of the breeding season.

DISTRIBUTION: Damp deciduous and mixed woodland with open, vegetated clearings; on plains and in mountainous regions up to the tree line; some winter in the Mediterranean, others remain in Central Europe.

FOOD: Worms, insects, larvae, spiders, millipedes and other ground-dwelling animals, plucked from the forest floor using its long beak ①.

TYPICAL FEATURES

When startled, the Woodcock flies away with clapping wing beats and with a characteristic zig-zagging flight, just above ground level. Red-brown rump is visible during flight ③.

NESTING: Nest is a flat hollow in the ground, lined with leaves; three–five cream-coloured eggs with rust-brown speckling; young are nidifugous.

Common Snipe
Gallinago gallinago

APPEARANCE: About 27cm-long wading bird with a very long, straight beak ①; back has brown flecking with whitish, lengthwise stripes; head has a contrasting light and dark striping; tail has white edges ②.

VOICE: Courtship call is a monotonous and clock-like 'tocka tocka tocka'.

DISTRIBUTION: In moors, marshes and silted areas on riverbanks and lake shores. It is Central Europe's most common snipe, but it is secretive and rarely seen; short-distance migrant.

TYPICAL FEATURES

When threatened, the Common Snipe first flattens itself motionless against the ground. Only at the last moment does it fly away in a zig-zag path, uttering its characteristic 'etsch etsch'.

FOOD: Worms, snails, small crabs, insects and insect larvae.

NESTING: Three–five olive-brown eggs with dark speckles; eggs laid in a hollow in the ground in dense vegetation. Young are nidifugous.

SIMILAR SPECIES: The smaller Jack Snipe (*Lymnocryptes minimus*) ④ has a shorter beak ③ and wide, yellow stripes on its back.

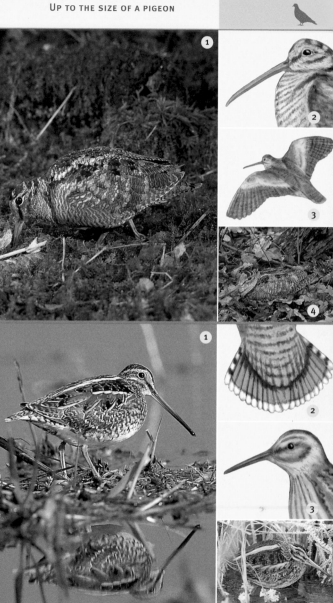

111

Common Redshank
Tringa totatus

APPEARANCE: About 28cm long, slim wading bird with long, bright red legs and a red base to its beak; plumage is grey-brown, in summer with dense, dark speckling ①, which is less pronounced in its winter plumage ③.

VOICE: Typical call is a two or three-syllable, melodious 'tew-ew' or 'tew-ew-ew'; alarm call is a loud 'chip chip'.

TYPICAL FEATURES

In flight, the broad, white rear wing edges are visible, as is the Redshank's white rump ②.

DISTRIBUTION: Nests in marshy fields, wet upland moors and bogs in coastal regions; occasionally found inland in Central Europe; outside of the breeding season, most common on mudflats; in passage also seen on riverbanks and lake shores; some winter on mudflats, others in Southern Europe.

FOOD: Insects, their larvae and earthworms, picked out from the mud ④; also tadpoles and small frogs.

NESTING: Nest on the ground, hidden in dense vegetation; three–five brown eggs with black speckles. Both parents raise the nidifugous chicks.

Ruff
Philomachus pugnax

APPEARANCE: Male is roughly 30cm long; breeding plumage has a featherless, yellow face, thick neck feathers and feathered tufts on the head, ranging in colour from white ③ to red-pink ① to black; female is only 20–25cm long, with unremarkable sand-colouring ②.

VOICE: In flight, an occasional 'cru' or 'geah' call, otherwise, the Ruff in mainly silent.

TYPICAL FEATURES

In June, the males begin to lose their thick neck plumage. In their winter plumage, they appear very similar to the female ②.

DISTRIBUTION: Mainly breeds in northern parts of Europe, on marshes, moors, heathland and dunes; some winter on the coasts of Western and Southern Europe, others in South Africa and in South Asia.

FOOD: Insects, worms and snails.

NESTING: Nest on the ground, well-hidden in vegetation; normally four eggs, with a heavy dark brown speckling; young are nidifugous.

OTHER: At the end of March, the males collect on traditional breeding grounds, where they perform ritualised show-fights using their wildly ruffled plumage ④, in order to attract a female.

Little Grebe
Tachybaptus ruficollis

APPEARANCE: 25cm long, with a round body; in summer its back is dark brown, almost black, and its cheeks, throat and front of the neck are a bright red-brown ①; the base of the beak has a bright, green-yellow spot ③; winter plumage is an unremarkable grey-brown ②.

VOICE: Call is a light, whistling 'fiit fiit' or a thin 'pe-ep'; during the courtship period the male and female often perform a long, vibrating trilling in duet.

TYPICAL FEATURES
When the Little Grebe feels threatened, it will submerge itself as far as the neck ④.

DISTRIBUTION: In summer, it can be found on ponds with dense bank vegetation, slow-flowing rivers and on silt marshes near lakes; in winter it is found on open water on rivers and lakes, as well as in town centres.

FOOD: Water insects, snails, small crustaceans and tadpoles, normally caught whilst diving.

NESTING: Nests on water (p21) in dense vegetation; usually two, but sometimes three clutches per year of four–six white eggs; young are nidifugous.

Water Rail
Rallus aquaticus

APPEARANCE: 22–28cm long; sides of the head, throat, breast and belly are a uniform slate-grey; flanks have black and white striping ①, underside of the tail is cream-white ②; beak and eyes are red; juvenile in its first year ④ has no grey, and brown beak and eyes.

VOICE: The Water Rail is a noisy bird and often utters a loud screech, or a soft grunting, moaning, groaning or murmuring. On spring evenings, the male courtship call resounds like a repetitive 'tuk tuk tuk tuk'.

TYPICAL FEATURES
Water Rails normally run away from danger on foot. If they fly for a short distance, their flight appears clumsy, with laborious wing beats and dangling legs ③.

DISTRIBUTION: Solitary bird, found in reeds on river and lake shores, but also near ponds surrounded with dense vegetation and in ditches; some winter in Central Europe, others in Southern Europe.

FOOD: Water insects, larvae, tadpoles, worms, snails.

NESTING: Nests in dense bank vegetation; six–ten brown speckled eggs.

115

Common Moorhen
Gallinula chloropus

APPEARANCE: Roughly 35cm long; beak is red with a yellow tip, and a bright red bill base and forehead above; plumage is black and dark, olive-brown ①, underside of the tail is snow-white ②; legs are yellow-green with long toes ③; chicks are black, with a blue-red head patterning; juvenile is a uniform brown ④.

VOICE: Sharp 'kurr'ukk' or 'kirrek' call; during courtship a prolonged 'kreck kreck' can be heard, mainly at night.

TYPICAL FEATURES
Moorhens constantly bob their head while swimming and frequently flash their tail, displaying its white underside ②.

DISTRIBUTION: On lakes, ponds and slow-running rivers with lush bank vegetation, on village and park ponds, even in town centres; common everywhere, apart from high mountain regions.

FOOD: Marsh and water plants, seeds and fruits, insects, worms and snails.

NESTING: Reed nest, well-hidden in bank vegetation; one–three clutches per year of five–ten cream-white eggs with dark brown spots.

Coot
Fulica atra

APPEARANCE: About 37cm long water bird with a round body and black plumage ① and a bright white forehead above a similarly white beak ②; chicks have a red and yellow head ③; juvenile plumage is grey-brown with a whitish face and neck ④.

VOICE: The female has a loud, barking 'kut' call; the male's call is a short, harsh 'piks' or a voiceless 'pt', similar to a popping champagne cork.

TYPICAL FEATURES
Coots have a noisy take-off from the surface of the water. They flap their wings loudly, and slap the surface of the water with their feet.

DISTRIBUTION: On lakes, ponds, reservoirs and slow-flowing rivers, also on small park ponds, even in town centres; very common.

FOOD: Will eat almost anything, from water plants, reed shoots and grass to small animals such as insects and snails; in towns, also eats bread and scraps.

NESTING: Large, flat reed nest, normally in shallow water and protected by bank vegetation; five–ten red-brown to black spotted eggs.

117

Common Tern

Sterna hirundo

APPEARANCE: 31–35cm long and slim, with a deeply forked tail and narrow wings ②; beak is red with a black tip; breeding plumage has a black head cap ①; on the winter plumage only the back of the head is black ③.

VOICE: A shrill, descending 'kee-yarr' or a sharp 'kit kit kirr'; alarm call is a rapid 'kekekekekek'.

DISTRIBUTION: Common breeding bird on flat sea coasts, also breeds inland near large lakes and wide rivers; winters in West Africa.

FOOD: Small fish, crustaceans, tadpoles, water insects.

TYPICAL FEATURES
When looking for food, the Common Tern flies just above the surface of the water. When it spots its prey, it dives like an arrow into the water.

NESTING: Nest is a shallow hollow in dunes, or on sand and gravel banks; two–three sand-beige to olive-green coloured eggs with brown speckles.

SIMILAR SPECIES: The Arctic Tern *(Sterna paradisaea)* ④ only lives on sea coasts and has a red beak without the black tip.

Grey Partridge

Perdix perdix

APPEARANCE: Roughly 30cm long, stocky fowl with a short tail and a brick-red face ①; male (on the right in the picture) has a black-brown patch on its grey underside; on the female this is only faintly visible ③; juvenile is brown with fine, light striping ④.

VOICE: When the Grey Partridge is startled, it emits a loud 'pitt pitt' or 'rep rep' call. The territorial call of the male is a grating 'karrwick'.

TYPICAL FEATURES
In flight, the Grey Partridge's red-brown steering feathers can be seen at the sides of the tail ②.

DISTRIBUTION: In dry, small enclosed meadows with hedgerows and scattered bushes. Due to increased use of land for agriculture, this type of habitat has been greatly reduced, resulting in a reduction in the numbers of Grey Partridge.

FOOD: Grasses, clover, wild herbs and seeds; also eats insects.

NESTING: Nest is a shallow hollow in the ground beneath dense undergrowth, thinly lined with grass and leaves; 10–20, sometimes as many as 24, olive-brown eggs; chicks are nidifugous, and are cared for by both parents.

119

Grey-headed Woodpecker
Picus canus

APPEARANCE: About 32cm-long woodpecker; head and neck are grey, the rest of the bird is moss-green, with a yellow rump ②; male ① has a bright red patch on the forehead; the female ③ has no such patch.

VOICE: In spring, a descending series of soft, whistling tones, becoming slower towards the end, a bit like a 'gugugugu-gu-gu-gu gu gu'; rarely heard outside the breeding season.

TYPICAL FEATURES

In order to mark its territory, the Grey-headed Woodpecker performs a regular drumming at two-second intervals.

DISTRIBUTION: Quite common breeding bird, but has become rarer in certain areas; found from the Alps to the southern edges of North European lowlands; in deciduous and mixed woodland, parks, orchards and cemeteries.

FOOD: Mainly ant pupae and ants, but also other insects, seeds, berries and other fruit; mainly searches for food on the ground.

NESTING: Nesting holes mainly in the trunks of deciduous trees ④, self-constructed or abandoned by a Green Woodpecker; seven–nine white eggs.

Green Woodpecker
Picus viridis

APPEARANCE: A good 25cm-long, moss-green woodpecker with black eye-mask and an extended, bright red cap on the crown ②; female has a black, moustache-like stripe ①; on the male ② this patch has a red centre; yellow rump; underside of the juvenile has dark, horizontal striping; back has white flecking ③.

VOICE: Both males and females mark their territory with a prolonged, resounding laugh, a bit like a 'kew-kew-kew'; drums very rarely.

TYPICAL FEATURES

The entrance holes to Green Woodpecker nests are round or slightly oval. They are normally located low down in ancient or old deciduous trees ④.

DISTRIBUTION: In deciduous and mixed woodland, field coppices, orchards, gardens and parks with established trees.

FOOD: Mainly ants and their pupae, extracted from their nests using its long, sticky tongue; also eats other insects, worms, snails and fruit. Green Woodpeckers search for food almost exclusively on the ground.

NESTING: Nesting holes, either self-constructed or re-used; five–eight white eggs.

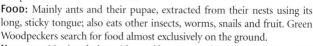

Fieldfare
Turdus pilaris

APPEARANCE: A good 25cm long; head and rump are light grey, back is red-brown, tail is black ②, underside has prominent black markings, and the flanks have characteristic arrow-shaped flecks ③.

VOICE: In flight, the Fieldfare often utters a loud 'chack-chack'; when the nest is disturbed it emits a croaky 'trrtrrtrr'. Song is a short chattering, often performed during flight.

DISTRIBUTION: In field coppices and on the edges of woods, in parks and gardens with abundant trees; in winter flocks of additional winter visitors arrive from North-east Europe.

FOOD: Worms, snails, insects, also berries ④ and fruit.

NESTING: Nests in colonies in trees ①; normally two clutches per year of four–six green-blue eggs with red patterning.

TYPICAL FEATURES
Both the Fieldfare and the Mistle Thrush have white patches under the wing which can be seen during flight. The two species can, however, be distinguished by their tails.

Mistle Thrush
Turdus viscivorus

APPEARANCE: About 27cm long; back is green-brown, underside cream-white to pale ochre-yellow, with dense black flecking ①; tail is brown with white corners ③.

VOICE: Flight call is a croaking 'trrrrr'; song consists of short, flute-like phrases, interrupted by long pauses; often heard singing during bad weather, for example in light rain, when other birds are silent.

DISTRIBUTION: In light woodland, preferably in dry conifer woodland, but also in large gardens and parks. Most migrate in winter to Western and Southern Europe, but some winter in Central Europe. Often seen in trees with large amounts of mistletoe.

TYPICAL FEATURES
In flight, a white patch can be seen on the underside of the wings ②. The flight of the Mistle Thrush has a distinct undulation.

FOOD: Worms, snails and insects, also berries; normally looks for food on open fields ④.

NESTING: Large nest, strengthened with mud, located high in trees; normally two clutches per year of four–six blue eggs, with red speckles.

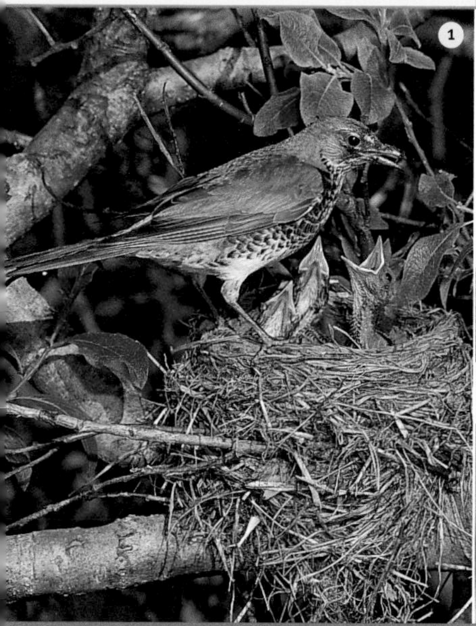
1

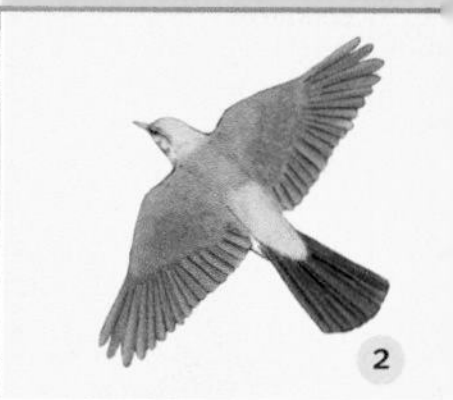
2

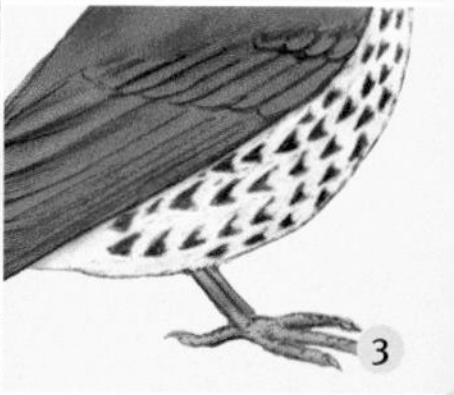
3

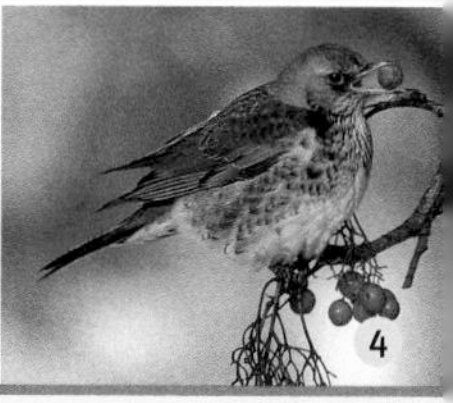
4

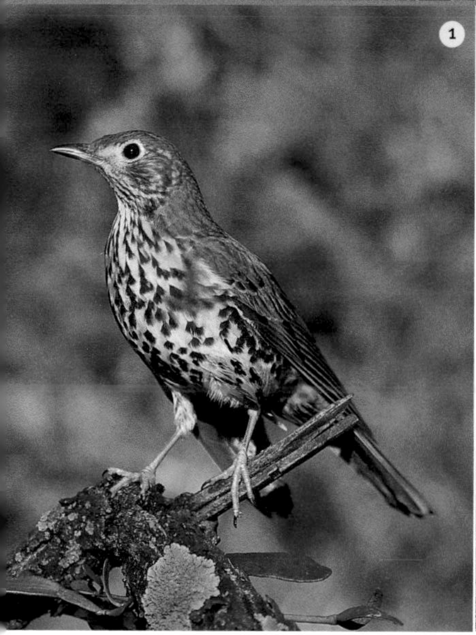
1

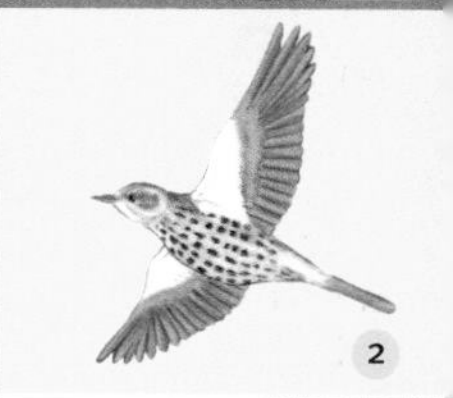
2

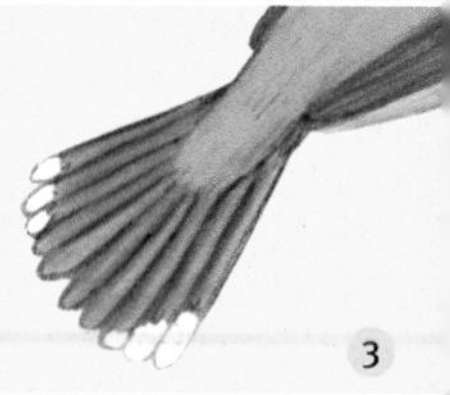
3

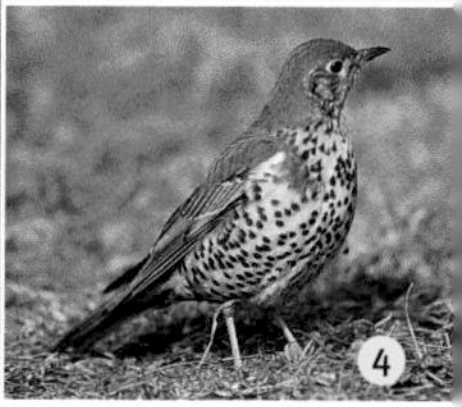
4

Collared Dove
Streptopelia decaocto

APPEARANCE: At 32cm long, the Collared Dove is only slightly smaller than the Common Street Pigeon (p126), but it is slimmer and has a longer tail ①. Its beige-brown plumage has a narrow black band around the neck, open at the front ②.

VOICE: In flight, often emits a nasal 'kwurr kwurr'; territorial song of the male is a monotone 'coo-cooo-coo coo-cooo-coo' with heavy emphasis on the middle syllable.

DISTRIBUTION: Originally from the plains of Southern Asia, but has been found in Europe since the 1930s; likes to live near humans in villages and towns.

TYPICAL FEATURES
Unlike the Turtle Dove, the Collared Dove is white-coloured ④, rather than blue-grey. Juvenile plumage is more grey than the adult plumage and lacks the collar ③.

FOOD: Seeds, green plant parts and buds, also cereals, fruits or bread.

NESTING: Flat nest of dry twigs and roots, normally located on the fork of a branch; two–four clutches per year; almost always two white eggs.

Turtle Dove
Streptopelia turtur

APPEARANCE: Only 26–28cm long; the rust-brown feathers on the back have a contrasting, black centre, giving the bird its characteristic patterning ①; on both sides of the neck there is a prominent black and white striped area ③; in juveniles ④ the feathers on the back do not have the dark centre.

VOICE: Territorial song of the male is a monotone, purring 'pooorr-pooorr-pooorr'.

DISTRIBUTION: Widespread and common in the Mediterranean; rarer in Central Europe and only found in lowland areas; on the edges

TYPICAL FEATURES
The Turtle Dove has a very quick but jerky flight. When seen from below, its fanned tail has a much narrower stripe along its edge ② than the Collared Dove.

of woods and forest meadows, field coppices and orchards, sometimes also in parks and gardens with deciduous trees; winters in sub-Saharan Africa.

FOOD: Plant and grass seeds; also pine nuts.

NESTING: Flat nest of twigs, normally quite low down in a tree or bush; normally two clutches per year of two white eggs; both parents build nest and care for young.

125

Common Street Pigeon
Columba livia f. domestica

APPEARANCE: About 33cm long; feral descendent of the Domestic Pigeon; great variety of colouring and patterns, from black to rust-brown to almost white; most are, however, 'pigeon grey' with a green or purple shimmer on the neck ① and two black wing-bars ②; eyes are orange.

VOICE: Loud cooing: 'coocu-curoo'.

TYPICAL FEATURES
Regardless of plumage colour, most Common Street Pigeons have a white or light grey rump.

DISTRIBUTION: In almost all European towns, mainly in cities, normally in large numbers; not only in parks and green spaces, but also on market squares and in areas with heavy traffic.

FOOD: Seeds, grains, buds, shoots, leaves, bread, scraps.

NESTING: Nests often on building ledges or balconies, in wall niches or under bridges; three–four clutches per year of two white eggs.

SIMILAR SPECIES: The Stock Dove *(Columba oenas)* ④ has less prominent wing-bars, and its eyes are dark ③ and not orange.

Eurasian Jackdaw
Corvus monedula

APPEARANCE: A relative of the crow, but only 33cm long; almost completely black, with a grey hood ①; beak is black; iris of the eyes is a whitish-grey ②; juvenile is initially brown ④, then black without the grey hood, but its eyes are still light grey ③.

VOICE: Call is a repeated, resounding 'tchack', and when endangered a high 'chiup'; song is a chatter, with miaowing and clicking tones, rarely heard.

TYPICAL FEATURES
The sociable Jackdaw often mixes with the Rook (p140) in winter, but it can be easily distinguished, even from a distance, by its grey neck plumage.

DISTRIBUTION: In light deciduous woodland and parks with woodpecker holes; also near churches, castles and ruins.

FOOD: Worms, snails, insects, fruits, seeds, occasionally mice and young birds; will eat anything and regularly searches dustbins.

NESTING: This sociable birds nests in colonies of various sizes. Very large nest in tree holes, cliff or wall niches; also nests in nesting boxes; four–six light blue, black speckled eggs.

127

Jay

Garrulus glandarius

APPEARANCE: The most distinguishing feature of this 34cm-long, pale rust-brown songbird is its light blue and black striped wing coverts ①. The Jay also has black moustache-like stripes either side of the beak ③.

VOICE: When disturbed, a loud, rattling 'scaaarg'; song consists of chattering, clicking and miaowing sounds.

TYPICAL FEATURES

In flight, the Jay's characteristic snow-white rump and white wing patches are clearly visible ②.

DISTRIBUTION: The Jay is a common inhabitant of deciduous and mixed woodland, but is also found in parks and large gardens with established trees. Commonly found in Central Europe all year round.

FOOD: Acorns, beechnuts, hazelnuts and insects; in spring also often eats eggs and chicks; hides tree fruits in cracks in bark, between roots and in the ground, to provide a winter store ④.

NESTING: Nest of twigs in dense branches of trees and large bushes; four–six blue-green to olive-brown, brown speckled eggs.

Nutcracker

Nucifraga caryocatactes

APPEARANCE: About 32cm long; head has a relatively long beak and dark brown crown ②; white, pearl-like flecking on a chocolate-brown plumage ③; wings and tail are black ①; tail also has white edges ④.

VOICE: Hoarse, rattling 'grrarr-grrarr' call; song is a quiet, melancholy chatter with much imitation of other birds; only rarely heard.

TYPICAL FEATURES

From a distance, the Nutcracker appears to be grey. It likes to perch on the very end of conifer branches.

DISTRIBUTION: In conifer and mixed woodland in low mountain ranges and the Alps, mainly in mountain pine forests. Only accidentally strays into British skies.

FOOD: Mainly conifer seeds (preferably Swiss Pine), but also hazelnuts and walnuts, berries and other fruits; in summer also eats insects and other small animals; like the Jay, creates winter stores, hiding nuts and tree seeds.

NESTING: Sturdy bowl-shaped nest, normally high in conifer trees, near the trunk; three–four pale, turquoise blue eggs with grey or brown speckling.

129

Hobby | Eurasian Hobby
Falco subbuteo

APPEARANCE: Male about 28cm, female 35cm long; slim, with very long, narrow wings ②; both male and female have a dark slate-grey back; whitish underside has dense flecking ①; characteristic orange-brown 'trousers' and rump, which are lacking on the juvenile bird ④.

VOICE: During the breeding season, often emits an incessant, rapid 'kew kew kew kew'.

TYPICAL FEATURES
The contrasting black and white face markings ③ of the Hobby can be seen from a distance.

DISTRIBUTION: Widely distributed throughout all Central European lowland, but not common in any location; needs open landscape in order to hunt, mainly meadows and moorland; occasionally also found near small human settlements; winters in Africa.

FOOD: The Hobby catches almost exclusively small birds and flying insects. It is extremely quick and agile when hunting in flight.

NESTING: The Hobby lays its two–four red-brown, heavily patterned eggs in abandoned Crow or Magpie nests, or the abandoned nests of other birds.

Common Kestrel
Falco tinnunculus

APPEARANCE: Male 33cm, female 39cm long; long, narrow wings and a similarly long tail with a black tip; male ④ has a light grey head and tail, and a brick-red back and wings; female ① and juvenile ③ are red-brown with pronounced flecking and striping.

VOICE: The call of the Kestrel is often heard and is, for example, a light 'keekeekeekee', or, on the nest, a whimpering 'vriiiih'.

TYPICAL FEATURES
Kestrels are normally seen hovering in the air ②, scanning the ground below for their prey.

DISTRIBUTION: Widespread from the sea coast to high mountains; breeds in field coppices and in Alpine cliff faces, quarries or church towers; not found in dense forests regions; always hunts over open land.

FOOD: Mainly mice, but also lizards, small birds, large insects.

NESTING: Like all falcons, the Kestrel does not construct its own nest, and instead lays its four–six yellow-white, densely red-brown speckled eggs in cliff crevices, holes in walls or abandoned Crows' nests.

131

Peregrine Falcon
Falco peregrinus

APPEARANCE: Male about 35cm, female up to 50cm long; back is slate-grey, underside is light with slight striped patterning ①; on the face there is a broad black tear-like stripe ④; talons are yellow ③; juvenile has a dark, grey-brown back, and a beige underside with tear-shaped flecks in length-wise rows ②.

VOICE: In the nest, a sorrowful 'geeah geeah'; when alarmed, a sharp 'kek-kek-kek'.

DISTRIBUTION: Found almost worldwide, but has become very rare in Central Europe; breeds on cliff faces and in river valleys in mountainous regions, but also on sea-cliffs; hunts in open countryside.

FOOD: Almost exclusively birds, caught in the air.

NESTING: Does not build a nest; a clutch of three–five densely brown speckled eggs is laid on a cliff ledge.

TYPICAL FEATURES
The Peregrine Falcon has a very characteristic hunting behaviour: during its spectacular hunting flights it can sometimes dive several hundred metres.

Sparrowhawk
Accipiter nisus

APPEARANCE: Male 28cm, female up to 38cm long; underside of the female ③ is light, but in the male ① has a rust-red colouring; both have tight, dark banding; juvenile plumage ④ has a faint rust-red patterning.

VOICE: During its circling courtship flight, emits a soft, ascending 'giu-giu' call; when disturbed near the nest and as a communication call between breeding partners, a long series of 'keekeekeekee' sounds.

DISTRIBUTION: In Central Europe it is very widespread, but only common in certain regions; inhabits conifer or mixed woodland, interspersed with open areas of land, hedgerows and copses.

FOOD: Over 90 per cent of its diet consists of birds, mostly caught in flight; occasionally also mice or bats.

NESTING: Nests in a high conifer tree, near to the trunk; five–seven brown speckled eggs (p23); it is usually the male who cares for the young.

TYPICAL FEATURES
The Sparrowhawk's characteristic flight profile has short, rounded wings and a narrow, straight tail, which is scarcely fanned, even when gliding ②.

Barn Owl

Tyto alba

APPEARANCE: 33–39cm long, very light-coloured owl with heart-shaped face patterning; back is gold-brown, interspersed with grey ②, breast and belly have variable colouring, ranging from white ① to sand-coloured to gold-brown ③; juveniles, as with all owls, have very similar plumage to adult birds ④.

TYPICAL FEATURES
The loud and croaking begging call of the female and young can often be heard near Barn Owl nests.

VOICE: During the breeding season, screeching and croaking sounds; 'song' of the male is a harsh, vibrating hiss.

DISTRIBUTION: Mainly in populated areas with few trees on the outskirts of villages; in Central Europe only found near humans.

FOOD: Predominantly field mice, but also other small mammals, small birds, frogs and large insects.

NESTING: On roof beams, in church towers, barns and other buildings; one–three clutches per year of two to 10 white eggs. Male provides food for the nesting female and, later on, the chicks.

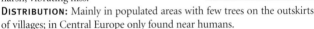

Short-eared Owl

Asio flammeus

APPEARANCE: 34–42cm-long owl with large flecking, yellow eyes ① and small, feathered ear tufts, which are raised when the bird is alarmed ②; at all other times, these ears are hidden in the head plumage; in flight, yellow patches can be seen on the wing primary feathers ④ and the wings also have a white rear edge.

TYPICAL FEATURES
Its yellow eyes are emphasised by a black rim. This rim is even visible on the completely beige ③ immature plumage of the juvenile.

VOICE: Male call is a yapping 'kfer-fek-fek'; female emits a 'warrr' sound; song of the male consists of an ascending eight to 12 syllable 'bu-bu-bu-bu…'.

DISTRIBUTION: In Central Europe, only a regular breeding bird in north-western lowland areas; otherwise it is found sporadically, depending on the food available; lives in open bogs and moorland with broad marshes.

FOOD: Predominantly voles, but also other small animals.

NESTING: Nest is a hollow in the ground, in dense vegetation, thinly lined with grass stalks; one–two clutches per year of four–eight white eggs.

1

2

3

4

1

2

3

4

Long-eared Owl
Asio otus

APPEARANCE: 35–37cm-long owl with yellow eyes and long, feathered ear tufts ①, which lie flat against the head when at rest ②; bark-coloured plumage provides good camouflage when perched in trees close to the trunk ④.

VOICE: Active at dawn, dusk and during the night. Its territorial song consists of a quiet, monotone series of 'ooo' calls.

DISTRIBUTION: Open countryside with small clumps of trees, coppices and open woodland; in winter occasionally also found in parks and gardens; not found in dense woodland.

TYPICAL FEATURES
When in flight, the Long-eared Owl's wide ochre-yellow and black wing patches are clearly visible ③. Unlike the Short-eared Owl (p134), wings have no white rear edge.

FOOD: Predominantly mice, but in winter also eats small birds; hunts in open countryside; in summer eats large insects.

NESTING: Three–eight white eggs, laid in abandoned Crow or Magpie nests or those of other birds of prey; the female incubates the eggs alone, and the male brings food.

Tawny Owl
Strix aluco

APPEARANCE: Roughly 38cm-long screech owl; basic plumage colour varies from red-brown ① to grey ③; two rows of white spots can be seen on the wings ②, crown is dark brown, edged with two light stripes.

VOICE: Territorial song of the male is a low 'hoo-hoo-hoo hooooo', heard from autumn onwards; the frequently heard, loud 'toowhit' call is normally uttered by the female.

DISTRIBUTION: The most common species of owl in Central Europe; found from lowland

TYPICAL FEATURES
The Tawny Owl has no feathered ear tufts like the smaller Long-eared Owl, and it can also be easily distinguished by its black eyes.

areas to Alpine forests; in light deciduous or mixed woodland, in parks, cemeteries and gardens with established deciduous trees, even in towns.

FOOD: Small mammals, especially mice, but also many birds; to a lesser extent also frogs and other small animals.

NESTING: Prefers to nest in tree holes ④, but also Crow's nests, cliff crevices, holes in walls, roofs or nesting boxes; three–five white eggs; female incubates the eggs, and the male provides the female and young with food.

137

Rock Ptarmigan
Lagopus mutus

APPEARANCE: About 35cm long; in winter both male and female are snow-white ④; in summer, back and breast have brown to black-brown marbling ②; in autumn and spring, varying degrees of chequered patterning ①.

VOICE: During its characteristic courtship flight, the male utters a wooden, croaking call, combined with loud wings beats. The male call is a 'corrr', while the female emits a 'kiia'.

TYPICAL FEATURES
Unlike the male, the female ③ does not have a black band stretching from the beak base and across the eyes.

DISTRIBUTION: Found year round in the Alps above the tree line, on cliff ledges and on stony Alpine meadows; in winter on steep slopes and southern cliff faces with little snow.

FOOD: Leaves and buds of Alpine herbaceous plants; berries, insects.

NESTING: Shabby nest in a hollow in the ground, well-hidden by dwarf bushes; four–nine eggs with red-brown to black speckles; young are nidifugous and are cared for by the mother until autumn.

Black Woodpecker
Dryocopus martius

APPEARANCE: 45–47cm-long, slim wood-pecker with coal-black plumage ①; male has bright red cap on the crown, stretching from the beak base to the back of the head ③, female only has a red patch behind the head ④; neck is curved during flight ②.

VOICE: Flight call is a multi-syllable 'croo-croo-croo', which can be heard over long distances; also emits a melancholy 'cleeooh' after landing; in spring utters an accelerating 'kwi-kwi-kwi…' call, and produces very powerful but relatively slow drumming.

TYPICAL FEATURES
Black Woodpecker holes have a long, oval entrance and usually are situated beneath the first branches of the trunk, normally at a height of 6–10m.

DISTRIBUTION: In expansive conifer and mixed woodland with many ancient trees; in mountainous regions up to the tree line.

FOOD: Insects, especially beetle larvae and large ants, living in decaying wood.

NESTING: Nesting holes are self-drilled, preferably in beech trunks, but also in pines; normally three–five white eggs.

139

Carrion Crow

Corvus corone corone

APPEARANCE: Roughly 47cm-long crow with a deep black, shiny plumage ①; beak very powerful, with a feathered base ②.

VOICE: Loud, screeching 'kaarr' or 'vaarrr' call; song is quiet and melancholy and rarely heard.

DISTRIBUTION: In Western Europe, as far east as parts of Eastern Germany and Central Austria; in open agricultural terrain.

FOOD: Insects, worms, snails, mice, frogs, eggs and juvenile birds, seeds and fruits, carrion and rubbish.

TYPICAL FEATURES

The Carrion Crow can be distinguished from the Rook by its feathered beak base and its thicker and more curved beak.

NESTING: Large, very sturdy nest, normally situated high in trees; three–six light blue to turquoise-coloured, brown speckled eggs.

OTHER: The Hooded Crow (*Corvus corone cornix*) is a sub-species of the Carrion Crow and is light grey on the neck, back and belly ④. Cross-breeds exist with small patches or flecks of grey ③.

Rook

Corvus frugilegus

APPEARANCE: Roughly 47cm long; plumage is completely black, with a blue or purple metal sheen ①; the beak base of the adult bird is bald and has a whitish, scabby appearance ③; in the juvenile bird, the beak base is still feathered ④; in flight, tail is slightly wedge-shaped ②.

VOICE: Croaky and deep 'kaah' or 'korr' call; song is chattering, with cawing and light metallic tones.

TYPICAL FEATURES

In winter, Rooks gathers in large numbers in the evening to roost. In the afternoon, they begin to fly in great flocks to these roosts.

DISTRIBUTION: Mainly in lowland areas; breeds in open cultivated areas with scattered field coppices; in winter also common in parks and towns; in Central Europe giant flocks of winter visitors arrive from more eastern breeding grounds.

FOOD: Insects and larvae, worms, snails, mice, also seeds and other plant parts, and human food scraps.

NESTING: Colonies of large nests made of twigs, normally high up in trees; three–five blue eggs with various shades of brown speckling.

141

Alpine Chough
Pyrrhocorax graculus

APPEARANCE: At 38cm, the Alpine Chough is larger and noticeably slimmer than the Eurasian Jackdaw (p126). Plumage is a shiny black ①; legs are red, beak is yellow and comparatively short ④.

VOICE: Sharp, shrill 'tjaa' or 'tschirr' call; communication call is a quiet and soft 'pruit'.

DISTRIBUTION: Inhabits high, Alpine regions up to an altitude of about 3,000m; in winter also found in valleys.

FOOD: Insects, worms, snails, fruits, carrion, rubbish.

TYPICAL FEATURES
The Alpine Chough is found in large numbers on much-climbed mountain peaks, near Alpine lodges and cable-car stations, where they are fed.

NESTING: Very large nests in cliff holes or crevices, occasionally also under roofs of buildings; four–five whitish eggs with dark speckling.

SIMILAR SPECIES: The very rare Chough (*Pyrrhocorax pyrrhocorax*) has a longer, thinner beak, which is red in adults ②, but yellow in juveniles ③.

Magpie
Pica pica

APPEARANCE: 44–48cm long, with a very long tail ①; black and white plumage, black tail and wing patterning with an intense metallic shine; juvenile ② has a shorter tail and no metallic colouring.

VOICE: Croaking, chacking call, a bit like a 'chackackack'; song of the male is chattering and quiet.

TYPICAL FEATURES
A Magpie can be easily identified in flight by its long tail, white wing patches, and white belly ③. From above, a white 'V' can be seen on the back ④.

DISTRIBUTION: In open countryside with field coppices and hedgerows, also in gardens, parks and cemeteries with tall trees, even in city centres; widespread throughout Europe and common in almost all lowland areas; in winter often seen in large groups at a common roost.

FOOD: Mainly snails, worms, large insects, eggs and juvenile birds, but also carrion, fruits, seeds and human rubbish.

NESTING: A large, enclosed nest of twigs, with a roof, normally constructed in the tree canopy; only one clutch per year of five–eight blue or green eggs.

143

Wood Pigeon
Columba palumbus

APPEARANCE: A good 40cm long; plumage grey, with red colouring on neck and breast ①; bright white patches on sides of neck ③ do not appear on juvenile birds in the first year ②.

VOICE: The territorial song of the male is a soft cooing like a 'coo cooooo coo cu-cu', normally stressed on the second syllable.

DISTRIBUTION: In deciduous and mixed woodland with surrounding meadows, fields and field coppices; also in town parks; very widespread and common in lowland areas; winters in regions with a mild climate, otherwise migrates to the Mediterranean to escape the cold.

TYPICAL FEATURES

Unlike Collared and Turtle Doves (p124), the Wood Pigeon has a black bar on the end of its tail ④. White wing markings are prominent in flight.

FOOD: Seeds, berries, tree fruits up to the size of acorns, grains, green plant parts, occasionally also insects and worms.

NESTING: Flat nest of twigs, normally high up in trees, sometimes in buildings; two–three clutches per year of two white eggs.

Black-headed Gull
Larus ridibundus

APPEARANCE: The breeding plumage of this 38–44cm-long gull has a chocolate-brown hood, leaving the neck free ①. During the winter, it only has a dark patch on the ear, and the rest of the head is white ④. On the juvenile plumage ③, the back and wings are brown.

VOICE: Often a croaky, aggressive-sounding 'kwerrr' and 'kree-aaa'; when competing for food, a light, repeated 'ke ke ke'.

DISTRIBUTION: Europe's most widespread gull, and by far the most common gull in inland areas; breeds on lake shores and small

TYPICAL FEATURES

In summer and winter, the Black-headed Gull can be recognised in flight by the wedge-shaped white patches on its primary wing feathers ②.

islands in lakes; outside the breeding season can be seen in all water habitats, and in winter it is also common in towns, parks and rubbish tips.

FOOD: Water insects, small fish, worms, carrion and rubbish.

NESTING: Normally nests in large colonies on a base of plant matter or on small islands in water; three–four eggs, the colour of which varies from rust-brown to olive to pale blue; all eggs have dark speckling.

145

Black-legged Kittiwake
Rissa tridactyla

APPEARANCE: 38–40cm long; back and upper surface of the wing are a uniform light grey, with black wing tips ④; remaining plumage is pure white during the breeding season; in winter plumage has grey areas around the ear and on the sides of the neck; juvenile ③ has a black neck band and a black patch behind the eyes.

VOICE: Very loud, bellowing 'kitiweek' call when near the nest; also a croaky 'geg geg'.

TYPICAL FEATURES
The Black-legged Kittiwake can be distinguished from the very similar Common Gull by its black legs ②.

DISTRIBUTION: Breeding colonies on steep sea cliffs on the northern Atlantic coastline; outside the breeding season can be seen on open lakes.

FOOD: Fish, small crabs, sea snails and plankton.

NESTING: Nests in colonies on narrow cliff ledges ①, normally on a base of earth and mud; normally two dark speckled eggs. For seven weeks, the nestlings are constantly watched by the adult birds, to avoid them falling from the nesting ledge.

Common Gull
Larus canus

APPEARANCE: 38–44cm long; back and upper wing surfaces are light grey, underside is white ①, wings tips are black with bright white flecks ②; in winter the head has pale grey-brown flecking, and the beak often has a dark ring ④; juvenile plumage is a dirty white with a light brown back.

VOICE: Call is high and shrill, a bit like a 'keetchoo keetchoo', 'kee yarr' or 'a-a-aiia'.

TYPICAL FEATURES
The Common Gull differs from the bigger but otherwise very similar Herring Gull (p172) by the yellow-green colouring of its legs ③.

DISTRIBUTION: Breeds in Central Europe on sea coasts, especially on the Baltic Sea; occasionally also found inland, especially in winter, often in groups with Black-headed Gulls.

FOOD: Insects, molluscs, small fish, mice, carrion, rubbish.

NESTING: Nests on firm ground, very close to water, either in low vegetation or on plain rocks; on the coast, always breeds in colonies; inland can also breed alone; normally three eggs, the colour of which varies from olive to rust-brown to sand-beige; eggs are always densely speckled.

147

Oystercatcher
Haematopus ostralegus

APPEARANCE: 40–45cm; head and back are black, belly is white, and beak, legs and eyes are bright red ①; the winter plumage has a white band across the throat and the sides of the neck ③; juvenile has a dark beak tip and brown eyes.

VOICE: Its loud 'kleep kleep' and a bellowing 'glee yarr' calls, uttered when in danger, cannot be missed.

TYPICAL FEATURES
The Oystercatcher's contrasting black and white colouring is unmistakable in flight ②.

DISTRIBUTION: This common coastal bird is found on flat, sandy or stony beaches of the North Sea and Baltic Sea; only found in small numbers in estuarial regions and inland waters; outside the breeding season often seen in large flocks.

FOOD: Mussels, shrimp, crabs, crustaceans, snails and mudworms.

NESTING: Nest ④ is a hollow on the beach, lined with pebbles, mussel shells or plant parts, in dunes or meadows with short grass; normally three eggs, which are a dirty yellow colour with black spotting.

Avocet
Recurvirostra avosetta

APPEARANCE: 42–46cm-long, slim coastal bird with long legs, black and white plumage and a thin, upward curved, black beak ①.

VOICE: Call is a soft, cheerful 'kloot kloot'; alarm call is a screeching 'quaatt', or an urgent, cackling 'kleep kleep kleep'.

TYPICAL FEATURES
In flight, the Avocet's black wing tips and completely white underside ② are fully visible. Its long legs extend far behind the tail.

DISTRIBUTION: On flat sea beaches, on lagoons and estuarial regions; common on the North Sea and Baltic Sea coasts; short-distance migrant, only a few winter on mudflats.

FOOD: Small crabs, worms, water insects and their larvae. When searching for food, the Avocet moves its beak through the water with side-to-side head movements ③.

NESTING: Nests in colonies; nests on the ground, often in the middle of a colony of gulls or other coastal birds; three–five clay-coloured eggs with black spots and speckles; young are nidifugous ④.

149

Black-tailed Godwit
Limosa limosa

APPEARANCE: Roughly 40cm-long wading bird; long legs and orange-yellow, black-tipped beak; breeding plumage has rust-red neck, which is more striking in the male ① than in the female; winter plumage of both male and female is grey-brown with light belly ②; in flight, broad white wing-stripe can be seen ④; juvenile has scale-like plumage on back ③.

TYPICAL FEATURES
During courtship, the male performs spectacular display flights, during which he plummets from the sky, or turns on his back.

VOICE: Frequent, somewhat nasal call, like a 'dididi', 'deeart' or 'weeka-weeka-weeka'.
DISTRIBUTION: Nests in damp fields, meadows or on heathland; outside the breeding season, seen in large numbers on flat shores; migrates to Southern Europe or North Africa for the winter.
FOOD: Earthworms, grasshoppers, beetles and snails; in shallow water also tadpoles and small crabs.
NESTING: Its three–five olive-brown, dark speckled eggs are laid in ground hollows, in high grass. Young are nidifugous.

Smew
Mergus albellus

APPEARANCE: Roughly 40cm long; male breeding plumage is snow-white with black markings ① and an extendable feathered crest ③; the male winter plumage is similar to the year-round female plumage, with a brown head ④; in flight, the bright, white wing-stripes are visible ②.

TYPICAL FEATURES
The Smew can be easily distinguished from other ducks by its darting flight with rapid wing beats.

VOICE: Harsh and rasping; very rarely heard outside the breeding season.
DISTRIBUTION: Breeds in Northern Eurasia, on the lakes and rivers of the taiga; a regular winter visitor in Central Europe, mainly in the Netherlands and northern Germany; occasionally found on lakes in the Alpine foothills.
FOOD: Small fish, water insects, small crabs and mussels.
NESTING: Nests in tree holes, generously lined with feathers; six–nine cream-white eggs; young leave the nest after 10 days.
OTHER: Smews are often found alongside the Common Goldeneye (p176). Cross-breeds of the two species have been known.

151

Common Teal
Anas crecca

APPEARANCE: 34–38cm long; male breeding plumage has a stripe with a metallic green shine which runs from its red-brown head and down the neck ①; light yellow rump; winter plumage (July to September) is similar to the plain brown of the female ②.

VOICE: During courtship, a light, bell-like 'shring shring', similar to the noise of a cricket; female has a high screeching call.

TYPICAL FEATURES

In flight, the black and green speculum with a white front edge is clearly visible on both the male and female.

DISTRIBUTION: On ponds and small, shallow lakes with lush bank vegetation.

FOOD: In summer mainly insect larvae, mussels and snails; in winter mainly water plants and seeds.

NESTING: Ground nest in bank vegetation; six to 10 yellow-grey eggs.

SIMILAR SPECIES: The male of the rarer Garganey (*Anas querquedula*) ④ has a sickle-shaped white head-stripe. The female Garganey ③ has a more contrasting head patterning than the female Common Teal.

Mandarin Duck
Aix galericulata

APPEARANCE: This duck is about 45cm long; the male breeding plumage ① has long, decorative feathers on the head and sail-like, raised, orange-coloured wing feathers; female ④ is mainly grey-brown; distinguishing markings are a white eye-ring and eye-stripe, and a bright white belly ②.

VOICE: Flight call of the drake is a whistling 'frrik'; the female emits a bellowing 'koow'.

TYPICAL FEATURES

In the winter plumage, the male and female look very similar, but the male has a bright red bill, while the female bill is brown ③.

DISTRIBUTION: Originated in far-east Asia; found in Europe as a park bird; today, in certain regions, it has become wild and breeds freely, especially on lakes surrounded by woodland, and rivers with densely vegetated banks.

FOOD: Seeds and tree fruits, also worms, snails, insects.

NESTING: Mandarin Ducks normally nest in holes in trees and near to water. Seven to 12 white eggs are laid in the hole with no lining. The newly hatched chicks jump to the ground from the hole and are led to the water by the mother.

153

Tufted Duck
Aythya fuligula

APPEARANCE: 40–47cm long, round, diving duck; male breeding plumage is contrasting black and white, with long tuft on neck that has a metallic shimmer ①; on winter plumage this tuft is much shorter ④ and the plumage is brown on the flanks; female is dark brown ②.

VOICE: During courtship, the male emits a chick-like, thin whistling; the female call is a harsh quacking 'arr arr arr'; rarely heard outside the courtship period.

TYPICAL FEATURES
Many, but not all, female Tufted Ducks have a white fleck around the base of the beak.

DISTRIBUTION: Breeds on large lakes and reservoirs; in winter, population numbers increase with the arrival of passage migrants and winter visitors, and they can often be found in large numbers on both still and flowing waterways, also in parks in the centre of towns.

FOOD: Mainly snails, small crustaceans, mussels and insect larvae; to a lesser extent, it also eats the seeds of water plants.

NESTING: Nests in bank vegetation or in reeds; eight to 11 pale green eggs.

Pochard
Aythya ferina

APPEARANCE: About 45cm long, diving duck; male has a red-brown head; beak is black with a broad light grey stripe ②; in the breeding plumage, the breast is black, back and wings are silver-grey ①; winter plumage ③ is similar to the grey-brown female ④, and can only be distinguished by the head colouring.

VOICE: The female call is a croaky and rasping 'kerr kerr'; the courtship call of the drake sounds a bit like a 'wiwwierr'.

TYPICAL FEATURES
The rapid flight of the Pochard is accompanied by an audible whistling, caused by the flapping wings.

DISTRIBUTION: Originally a breeding bird of Eastern Europe, but has begun to breed further west over recent centuries; during breeding season it is mainly seen on shallow lakes surrounded by reeds; during the rest of the year, it can be seen on all waterways, and also on park lakes in towns.

FOOD: Water plants, seeds, insect larvae, worms, snails.

NESTING: Nest hidden in bank vegetation; five to 11 pale grey-green eggs.

OTHER: Pochard and Tufted Duck cross-breeds can occur, in which case they exhibit the characteristics of both parent birds.

155

Golden Eagle
Aquila chrysaetos

APPEARANCE: 75–90cm long, wingspan can exceed 2.2m; female somewhat larger than the male; powerful hooked beak with a black tip, prominent brows ④; both male and female adult birds are a uniform dark brown, and with age the head and nape become gold-brown ①.

VOICE: The Golden Eagle is almost never heard. Its call is, however, a 'hyarr' or a loud 'jick jick', similar to the call of a Buzzard.

TYPICAL FEATURES
Juvenile Golden Eagles have white wing markings ② and a white tail with a broad, black tip ③ until their fifth year.

DISTRIBUTION: The Golden Eagle is only be found in Scotland, where it has been reintroduced. When hunting in winter, it can fly great distances up and down the valleys and over the mountains.

FOOD: Mammals up to the size of a young chamois or deer (mainly marmots), birds up to the size of a Capercaillie, also carrion.

NESTING: Large nest, called an eyrie, on sheer cliff-faces, normally used for several consecutive years; normally two eggs per year, heavily speckled.

White-tailed Eagle
Haliaeetus albicilla

APPEARANCE: Male is about 80cm, female 90cm long, wingspan up to 2.45m; in adult birds the head and neck are a lighter brown than the rest of the plumage ①; tail is white and wedge-shaped ④; juvenile has a brown beak and tail ②.

VOICE: White-tailed Eagles are heard almost exclusively during the courtship period; the male emits a piercing 'krick-rick-rick-rick', and the female replies with a deeper 'ra-rack-rack'.

TYPICAL FEATURES
White-tailed Eagles can be easily identified by their powerful, yellow beak ③. Their legs are also a bright yellow colour.

DISTRIBUTION: Both on the coast and on large lakes and rivers with abundant fish stock; quite a rare bird in Central Europe, mainly in north-east regions; young White-tailed Eagles can wander far afield and have been seen on lakes in southern Germany.

FOOD: Mainly large fish, and waterfowl up to the size of a goose; sometimes also mammals, frequently eats carrion.

NESTING: Extensive eyries in tall trees on lake shores; on the coast, also on cliffs; one–three white eggs.

1

2

3

4

1

2

3

4

157

Osprey
Pandion haliaetus

APPEARANCE: 55–70cm long and slim, with gull-like, narrow wings ②; back is dark brown, underside is white ④; head is white with a black-brown mask ③; nape feathers often ruffled when perched.

TYPICAL FEATURES
From a distance, the underside of a flying Osprey looks mainly white ②.

VOICE: When near the nest, a frequent, short series of calls, descending in pitch at the end, a bit like a 'tchip-tchip-tchip-tchip-tchup-tchup-tchop-tchop'.

DISTRIBUTION: Breeds in Scandinavia and Russia, but recently re-introduced to Salisbury Plain. Otherwise extremely rare in Great Britain. Migrates to sub-Saharan Africa for the winter.

FOOD: Exclusively fish, caught out of the water in its strong talons after a diving flight ①.

NESTING: Its impressive nest (p21) is constructed in a tall tree and is often used for several years; one clutch per year of two–four brown speckled eggs.

Marsh Harrier
Circus aeruginosus

APPEARANCE: 48–55cm long; female is a chocolate-brown with a cream-coloured top of the head ①, and dark brown lower wing surfaces ③. The male plumage contains several colours with brown, light grey, black and beige areas ④; underside of the wings are mainly white with black tips ②.

TYPICAL FEATURES
When searching for prey, the Marsh Harrier flies just above the ground. Its wing beats appear slow and form a 'V'-shape.

VOICE: During the courtship flight, the male emits a screeching 'kewaerr' call; alarm call is a sharp 'keke-ke-ke'.

DISTRIBUTION: Only found in lowland areas, but not common anywhere; on lakes and rivers with broad reed banks; hunting flights over meadows and fields near water; winters in the Mediterranean or sub-Saharan Africa.

FOOD: Frogs, waterfowl up to the size of a Coot, small rodents, also often eats bird eggs and nestlings.

NESTING: Nest is on the ground, hidden in dense bank vegetation, and often also in reed banks floating on the water surface; three–six white eggs.

159

Common Buzzard

Buteo buteo

APPEARANCE: 50–55cm long, wings are broad, with a wingspan of about 1.20m; relatively short, often widely fanned tail; plumage on the back is mostly a uniform brown, but the underside has variable colouring, from black-brown, to dense brown flecking ①, to almost white ④.

VOICE: A high, mewing 'kiew'.

DISTRIBUTION: Along with the Kestrel, the Buzzard is Central Europe's most widespread and common bird of prey; nests on the edges of woodland, hunts over open, cultivated land, above fields and meadows.

FOOD: Small, ground-dwelling animals up to the size of young rabbits; main food is field mice, but also frogs, lizards, juvenile birds and even insects and earthworms; also eats carrion.

NESTING: Nest is high in trees, newly built each year; two–three dark speckled eggs; eggs incubated mainly by the female; the male provides food.

TYPICAL FEATURES
The Common Buzzard can be distinguished from the Honey Buzzard by its dark eyes ② and yellow feet ③.

Honey Buzzard

Pernis apivorus

APPEARANCE: Similar size and plumage to the Common Buzzard, but slimmer, with a smaller head ① and a longer tail; eyes are a bright yellow ②.

VOICE: When patrolling its nesting territory, a melancholy, three-syllable call, a bit like a 'hood-lew-hooh'; when agitated, a light 'kikikiki…'.

DISTRIBUTION: In almost all wooded landscapes; rarer than the Common Buzzard; nests on woodland edges, hunts for food over open countryside; migrates to tropical Africa for the winter.

FOOD: Mainly wasp, bee and Bumblebee larvae and pupae, dug out of nests in the ground; also eats other insects; on occasion also frogs, lizards or mice.

NESTING: Nests in high trees, regularly lined with fresh twigs with leaves ④; normally two chicks.

TYPICAL FEATURES
The tail of the Common Buzzard may have eight to 12 tight, dark tail-bands, but the Honey Buzzard can be identified because it has just three dark stripes on the tail ③.

161

Red Kite
Milvus milvus

APPEARANCE: 60–66cm-long bird of prey with a relatively long, forked tail; head is light grey ④, rest of the plumage is mainly red-brown ①; in flight, contrasting white wing flecks can be seen ②.

VOICE: Call on the nest sounds like a wail, a bit like a 'hya-hee-hee-hee-hya'.

DISTRIBUTION: In varied landscapes with established deciduous woodland, open fields and meadows, and water courses; winters in the Mediterranean, but also increasingly in Central Europe.

TYPICAL FEATURES

The relatively long tail of the Red Kite is quite deeply forked in the adult bird ③, but in the juvenile less so.

FOOD: Mainly carrion, but sometimes also catches mammals and birds up to the size of a hare or a chicken.

NESTING: Nest located in the canopy of tall trees, and is lined with grass and leaves, and often also paper, material and plastic; two–three brown speckled eggs; female incubates alone; the male provides food.

Black Kite
Milvus migrans

APPEARANCE: At 55–60cm long, it is some-what smaller than the Red Kite; plumage is obviously darker; head is grey-brown ②, and only slightly lighter than the black-brown body ①; tail only slightly forked ③, and when fanned, it appears perfectly straight ④.

VOICE: During the breeding period, the call is a shrill whining, a bit like a 'fiu-hihihihi…'; otherwise a 'hyaerr' call, similar to the Buzzard.

DISTRIBUTION: In wooded landscapes with many lakes and open spaces; widespread in Central Europe, but not found in Britain; winters in Africa.

TYPICAL FEATURES

It is common to see the Black Kite flying at low altitude along lake or river shores. It is searching for dead fish, which it then skillfully plucks from the water.

FOOD: Mainly dead or ill fish, but also ill or injured mammals and birds; also carrion; only rarely does it catch its own prey.

NESTING: Nests high in coniferous or deciduous trees, often lined with plastic or paper waste; two–three white, brown speckled eggs.

163

Goshawk
Accipiter gentilis

APPEARANCE: Female up to 60cm long; male about a third smaller; head relatively flat, with yellow eyes ③; plumage on back is grey-brown, and whitish on breast and belly with tight, horizontal bands ①; tail is long and narrow, wings appear rounded in flight profile ②; juvenile underside has dark brown flecks ④.

VOICE: Most common call is a light, prolonged 'hiiiah'; also a quick series of 'kek kek kek' sounds, or a short 'juk'.

DISTRIBUTION: Found throughout Europe; preferring well-lit coniferous and mixed woodland, sometimes also in wooded areas on the outskirts of towns.

TYPICAL FEATURES
Unlike the Sparrow-hawk, the Goshawk only circles in spring. Its flight consists of periods of rapid wing beats, followed by short gliding phases.

FOOD: Mainly birds and mammals, from the size of mice and small songbirds to the size of Capercaillies and hares.

NESTING: Nests in the canopy of tall trees on the forest edge; two–five pure white eggs.

Eagle Owl
Bubo bubo

APPEARANCE: Up to 70cm long, bulky build; eye colour ranges from yellow to orange-red; long, feathered ear tufts ②; plumage has dense brown flecking ①; feet are feathered up to the talons ③.

VOICE: Male has a deep, far-carrying 'buho' call; female call is somewhat higher and has two distinct syllables: 'u-huu'; mainly heard in autumn and just before spring; female also often emits a harsh 'shraae' or a sharp 'greck'.

TYPICAL FEATURES
In flight, the short, broad tail and particularly broad wings are characteristic of the Eagle Owl ④.

DISTRIBUTION: Mainly in the high mountain ranges; in regions with cliffs and woodland; also in lowland areas in river valleys with steep gorges or in large quarries.

FOOD: Mammals from the size of a mouse to that of a hare, also birds up to the size of a Capercaillie.

NESTING: The two–five white eggs are laid on a cliff ledge with no lining, sometimes also simply on the ground. The female incubates the eggs alone, and the male provides food for the family.

165

Capercaillie
Tetrao urogallus

APPEARANCE: Cock is about 85cm long, plumage is black-brown, with a green sheen on the breast; above the eyes there is a red patch of skin ③; hens ④ are only about 60cm long and have brown, well-camouflaged plumage with a rust-red breast patch and tail; unlike the Black Grouse, the tail is rounded during flight ②.

VOICE: Courtship song of the cock consists of clicking, accelerating sounds, ending with a loud 'pop', followed by some slurred tones.

DISTRIBUTION: In the Highlands of Scotland and in Europe in the mountain ranges.

TYPICAL FEATURES
During courtship, the male Capercaillie stands on the ground or a low branch with its fanned tail raised ① performing its characteristic 'song'.

FOOD: Herbaceous plants, berries; ant pupae fed to the young; in winter mainly spruce or pine seeds.

NESTING: Nests on the ground, often under low-hanging branches; five to 12 yellow-brown, densely speckled eggs; young are nidifugous.

Black Grouse
Tetrao tetrix

APPEARANCE: About 40–55cm long; male bird ① has blue-black shiny plumage, thick, bright-red eyebrow bulges and extended, lyre-shaped tail feathers ③; female bird ④ is plain brown, and has a slightly forked tail in flight.

VOICE: The courtship song of the male is a far-carrying rolling or gurgling. The female has a nasal cackling.

DISTRIBUTION: In Scotland and mountainous regions of Europe around the tree line; very rare in lowland areas, mainly on moors and heathland.

TYPICAL FEATURES
In flight, the Black Grouse displays broad white wing-stripes, as well as its curved tail ②.

FOOD: Buds of deciduous and coniferous trees, leaves and fruits of low berry bushes; feeds the young on insects.

NESTING: Nest on the ground, well-hidden in vegetation; seven to 10 speckled eggs.

OTHER: The male birds spend almost the whole year in the same courtship ground, where they court with pricked tails, heavy wing beats and fluttering leaps.

167

Ring-necked Pheasant
Phasianus colchicus

APPEARANCE: About the size of a domestic chicken, and including the tail, about 90cm long; male bird brightly coloured, mainly copper-red with dark green, shimmering head ①, often with white ring about the neck, and always with a red skin flap on the face ②; hen ④ is earth-brown with dark flecking; juvenile is similar to the hen, but with a shorter tail.

TYPICAL FEATURES
Both male and female birds have a long, pointed tail, making it easy to identify both in flight ③ and on the ground.

VOICE: Territorial call of the male is a bellowing 'kok kok kok', accompanied by audible wing claps; alarm call of the male bird is a hoarse, cawing 'ach-ach', the hen emits a sharp 'tsik-tsik'.

DISTRIBUTION: Originates from South-east Asia, but introduced to Europe centuries ago as a game bird; today is widespread throughout Europe in cultivated areas with fields and tree thickets; common in many places.

FOOD: Green plant parts, seeds, berries, cereals, agricultural crops.

NESTING: Nests on the ground, well hidden in vegetation; produces eight to 12 olive-brown eggs.

Raven
Corvus corax

APPEARANCE: With a body-length of 65cm, the Raven is the largest songbird in the world. Its 'raven-black' plumage has a metallic-blue shimmer ①. Its legs ③, eyes and noticeably long beak ② are also black.

TYPICAL FEATURES
The wedge-shaped tail is very characteristic of the flight profile of the Raven ④. Ravens are often seen gliding in pairs on air currents.

VOICE: In flight, a deep and loud 'kronk' or 'kroa', sometimes also a metallic 'klong' or a crow-like 'warrr'; its song is soft and chattering, but is rarely heard.

DISTRIBUTION: Found in a very wide variety of habitats, from coastal regions to woodland areas to high mountain ranges; in Central Europe it is mainly found in the Alps, and in northern areas.

FOOD: Small animals, for example insects, worms and snails, small vertebrates, carrion, human rubbish, but also seeds and fruits.

NESTING: Very sturdy nest of twigs in the canopy of a tall tree or in a crack in a cliff-face; three–six greenish eggs with dark speckles and patterning.

169

White Stork
Ciconia ciconia

APPEARANCE: With an extended neck, the White Stork is 1m long; wingspan about 2m; black primary feathers on otherwise white plumage; beak and legs are bright red ①; in juvenile storks ③ legs are black-brown.

VOICE: The White Stork does not have a call and instead communicates by clapping its beak ④. Nestlings emit a grunting, belching and whimpering sound.

TYPICAL FEATURES
In flight, the White Stork can be easily identified by its long, outstretched neck and long, trailing legs ②.

DISTRIBUTION: Open cultivated land with meadows and marshland; extremely rare in Britain. Winters in the south of Africa.

FOOD: All types of small animal, caught on the ground or in shallow water, for example, mice, frogs, fish, snakes and the young of ground-nesting birds; young are fed on earthworms.

NESTING: Large nest of twigs on rooftops, chimney stacks or electricity pylons, but also in trees; three–five white eggs.

Black Stork
Ciconia nigra

APPEARANCE: Body length is just under 1m; wingspan 1.7–2m; black, metallic-green and purple shimmering plumage ①, only the belly is white, with an abrupt change of colour on the breast ④; legs, beak and eye-ring are a bright red ③; juvenile ② is a dark grey-brown with olive-brown beak and legs.

VOICE: In flight, a melodic 'ouuu-o'; on the nest a soft 'hi-liii-hi-liii'; when agitated a light 'fiiieh', that develops into a short hiss; unlike the White Stork, it rarely emits a beak clap.

TYPICAL FEATURES
At the beginning of the breeding season, Black Storks glide for hours, high above their breeding territory.

DISTRIBUTION: In undisturbed deciduous and coniferous woodland with abundant streams and ponds. Extremely rare in Britain, but has recently begun to increase in numbers in woodland areas of Europe that are not used for forestry; migrates to East Africa for the winter.

FOOD: Mainly fish, frogs and salamanders.

NESTING: Large nest of twigs in tall trees; nest is used year long, and further extended each year; three–five white eggs.

171

Grey Heron
Ardea cinerea

APPEARANCE: 90–100cm-long, mainly grey-coloured heron with a powerful, dark yellow beak; broad, black eyebrow-stripe, which extends behind the head in long, thin, decorative feathers ②; juvenile has a black crown; at rest, the neck normally forms a 'S'-shape, so that the head seems to grow out of the shoulders ③.

VOICE: Loud, croaky call, a bit like a 'kraak' or 'frank'; on take-off, often emits a short, two-syllable 'kra-ik'.

TYPICAL FEATURES
In flight, the Grey Heron holds its neck in a 'S'-shape. Its long legs trail far behind the body ④.

DISTRIBUTION: Searches for food by water with densely vegetated banks ① as well as in marshes and bogs; breeding colonies are often located away from water, in small woods.

FOOD: Mainly fish, also frogs, lizards, mice, insects.

NESTING: Large twig nest in colonies, normally in tall trees, sometimes also in reeds; nests often used year round; three–five pale green eggs.

Herring Gull
Larus argentatus

APPEARANCE: 55–67cm long; breeding plumage is snow-white with silver-grey wing coverts and black and white wing tips ①; in the winter plumage, the head is striped with grey; juvenile plumage has fine, brown flecks until the fourth year ④.

VOICE: Howling or mewing 'kee-owk-kyowk-kyowk', often repeated several times; also a mewing 'kya kya kya keeyow'; alarm call at the nest is an 'ag-ag-ag'.

TYPICAL FEATURES
Characteristic features of the Herring Gull are a yellow beak with a red spot on the lower beak ③, yellow eyes and flesh-coloured legs ②.

DISTRIBUTION: The most common large gull in Northern Europe; breeds on the North Sea coast in tens of thousands; often breeds on waterways near the coast, and on large lakes and rivers further inland; in winter often found in ports and harbours, further inland near abattoirs and rubbish tips.

FOOD: Fish and other seafood, eggs, nestlings, carrion, human scraps.

NESTING: Ground nest, thinly lined with plant material, sometimes on bare cliff ledges; normally two–three green to olive-brown eggs.

173

Cormorant
Phalacrocorax carbo

APPEARANCE: 80–100cm long; plumage is mainly black; breeding plumage has a metallic green and bronze-coloured sheen; the head and neck have differing amounts of white colouring ①; on the hips there is a white fleck ②; beak tip has a sharp hook.

VOICE: Croaky 'chro-chro-chro' or 'krao', only heard at nest.

DISTRIBUTION: On sea coasts and on rivers and lakes with abundant fish stocks; numbers are increasing on inland waterways both as a breeding bird, and as a passage migrant or winter visitor.

TYPICAL FEATURES
When swimming, the Cormorant sits very low in the water, with its head and beak held upwards at a steep angle ③.

FOOD: Mainly fish, mostly between 10 and 20cm long.

NESTING: Twig nest in colonies in tall trees; on sea-cliffs; three–four light blue eggs with chalky coating; young are nidicolous.

OTHER: Cormorants often sit for hours at a time on an open perch with wings outstretched, in order to dry their wet feathers ④.

Great Crested Grebe
Podiceps cristatus

APPEARANCE: 45–50cm-long waterbird, with a long neck and grey-brown plumage; in summer has a rust-red and black feathered crest on its head, which can be raised to form 'ears' and 'sideburns' ③; in the winter plumage, this head decoration is absent or less conspicuous ②.

VOICE: Call is a croaky 'grook grook'; during courtship in spring, it also emits a rasping 'arrrr' or a voiceless 'k'pk'p'; at nest, a prolonged 'kweeah ah'.

TYPICAL FEATURES
During courtship, the partners perform a long, synchronised 'dance', with head-shaking, and extension of the neck ④.

DISTRIBUTION: On lakes and large ponds with extensive reed banks, also on slow-flowing rivers; in winter it is often joined by new arrivals from northerly regions.

FOOD: Small fish and other water animals, normally caught whilst diving; also insects and small crabs.

NESTING: Floating nest ①, normally hidden in reeds; four–five white eggs. The young can swim on their first day, but are normally transported on the backs of the adult birds.

175

Common Merganser
Mergus merganser

APPEARANCE: A 60cm-long swimming bird with a narrow, red beak; the male breeding plumage is black and white with a green shimmering head ①; the winter plumage has a red-brown head ③, similar to the female bird, which has a lighter grey back ④.

VOICE: The male courtship call is a frog-like 'kerr korr-kerr kerr'; female communicates with her young with a deep 'kro kro' call.

TYPICAL FEATURES
The beak of the Common Merganser has a sharp hook and fine sawing teeth along the edge ②.

DISTRIBUTION: On rivers and lakes with clear water and established trees on banks; in autumn and winter additional winter visitors arrive from more northern breeding grounds.

FOOD: Mainly fish, normally smaller than 10cm.

NESTING: Nests in tree holes, and holes in cliffs or walls; female incubates the eggs alone. The eight to 12 chicks jump out of the hole after one–two days and are led to the water by the mother.

Common Goldeneye
Bucephala clangula

APPEARANCE: Just under 50cm-long diving duck with a large head and yellow eyes; the male breeding plumage is black and white, with a round, white spot on the wing (①, right); male winter plumage is similar to the female ④, with chocolate-brown head, grey body, and yellow-tipped beak ③.

VOICE: During courtship, the males emit a screeching 'zeee ZEEE', during which the head is tilted far back at the nape ② and then thrust forward. Female emits a rasping 'quarr quarr' in flight.

TYPICAL FEATURES
The Common Goldeneye flies with very rapid wing beats, creating an audible whistling sound.

DISTRIBUTION: Only breeds in scattered locations in Central Europe, but is a winter visitor to British coastal regions, inland waterways and lakes.

FOOD: Water insects and their larvae, snails, small crabs.

NESTING: Does not nest in Britain. Nests normally in Black Woodpecker holes near water; six to 11 green eggs; chicks fall to the ground from the nest.

177

Mallard
Anas platyrhynchos

APPEARANCE: 55–60cm-long duck ②; the male breeding plumage has a shimmering green head and white neck-ring; the female is inconspicuous with brown flecking ①; both male and female birds have a blue-violet speculum ③.

VOICE: The male calls a croaky 'raarb raarb'; the female emits a loud quacking 'waaak-wak-wak-wak…'.

DISTRIBUTION: From the coast to mountain regions, on all types of standing water, and on slow-flowing rivers, in rural areas as well as in cities; the most widespread and common duck in Britain.

FOOD: Mainly water plants and small water animals and insects.

NESTING: Nest on the ground, normally near water, but sometimes at some distance; normally seven to 11 light brown-green eggs (p23); incubation and care of the nidifugous young is exclusively the role of the female.

> **TYPICAL FEATURES**
> In its winter plumage (June to September) the male has a uniform yellow bill ④, which distinguishes it from the dark-billed female bird.

Red-crested Pochard
Netta rufina

APPEARANCE: This diving duck is about 55cm long; the male breeding plumage has a thick, fox-red head ①; in its winter plumage, the male looks very similar to the female (④, left), with a dark brown crown and light grey cheeks; the beak on the male is always bright red ③, but on the female it is dark with an orange tip.

VOICE: When agitated, the male emits a croaky 'bat'; during the courtship period, a two-syllable 'ba-icks'; female emits a rasping 'korrrr', in flight, a 'wa-wa-wa'.

DISTRIBUTION: Mainly in Asia with a very small breeding population in Europe; on shallow lakes and ponds with plentiful food and lush bank vegetation; mainly winters in the Mediterranean.

FOOD: Mainly water plants, sometimes also small water animals.

NESTING: Nest hollow always located near water, well-hidden in vegetation and lined with stalks, feathers and down; six to 12 yellow-grey eggs; female incubates the eggs and cares for the young alone.

> **TYPICAL FEATURES**
> During flight, a prominent broad, white wing-stripe can be seen on both male and female birds ②.

Common Eider
Somateria mollissima

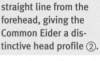

APPEARANCE: 50–70cm-long, heavy diving duck found on the sea coast; male breeding plumage is black and white ① with olive-green nape patch ②; the summer eclipse plumage is almost black, and only the front of the wings are white ③; females ④ and juveniles up to the third year have inconspicuous brown flecking.

VOICE: During the courtship period the male emits a resounding 'uhuo'; the female emits a rasping 'korrr-r'.

TYPICAL FEATURES
The beak is wedge-shaped, and runs in a straight line from the forehead, giving the Common Eider a distinctive head profile ②.

DISTRIBUTION: Breeds on Northern Atlantic coastlines, and sometimes also on the North Sea coast; during the moulting period and then as a winter visitor, large numbers can be found on the North Sea and Baltic Sea; in winter can occasionally be found inland.

FOOD: Small mussels, snails, crabs and water insects.

NESTING: Nests in a ground hollow, often in open countryside, with thick, down lining; four–nine pale, olive-brown eggs.

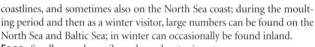

Red-billed Shelduck
Common Shelduck | *Tadorna tadorna*

APPEARANCE: 60–70cm long; plumage is black and white with broad, rust-brown band around the front of the body ①; female (①, right) often has white flecks on the face, and its red beak has no hump at the base; on the male the beak has a hump that is greatly swollen in spring ③; juveniles have a dark brown cap ④.

VOICE: Male emits a light whistling; the female has a rapid, almost cackling 'ak-ak-ak…'.

TYPICAL FEATURES
When dabbling ②, the Red-billed Shelduck reveals its rust-brown rump and black belly.

DISTRIBUTION: Flat, sandy beaches on the North Sea and Baltic Sea; sometimes on river estuaries and coastal lakes.

FOOD: Mud snails, annelids, small crabs, insect larvae.

NESTING: Nests in holes, e.g. in rabbit burrows, holes in the ground in dykes or bank vegetation, under rocks; eight to 12 yellow-white eggs.

OTHER: In July–August, tens of thousands of Red-billed Shelduck migrate from Europe to traditional resting places on the North Sea coasts, in order to change their plumage. From October they return to their breeding grounds.

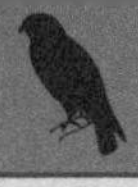

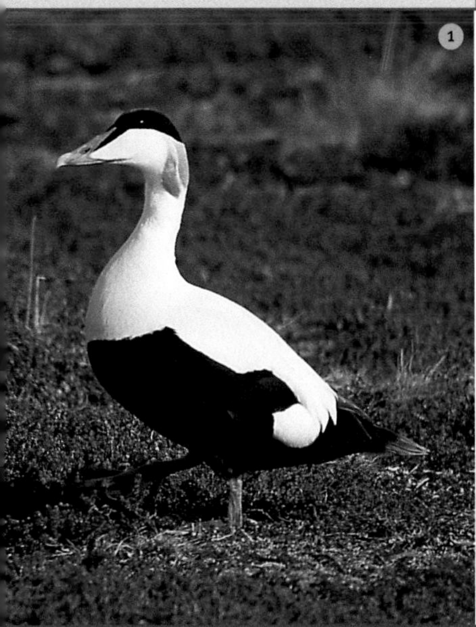
1

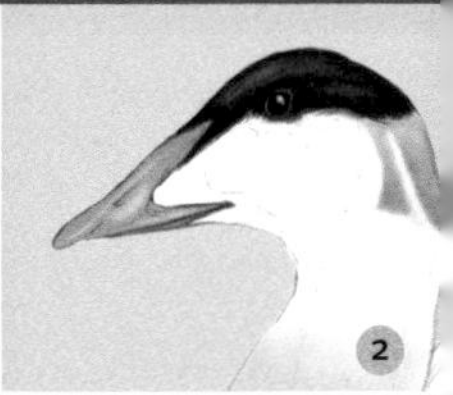
2

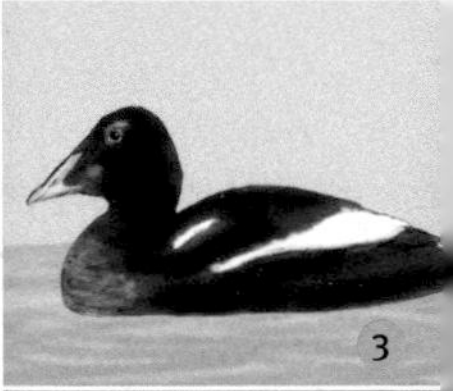
3

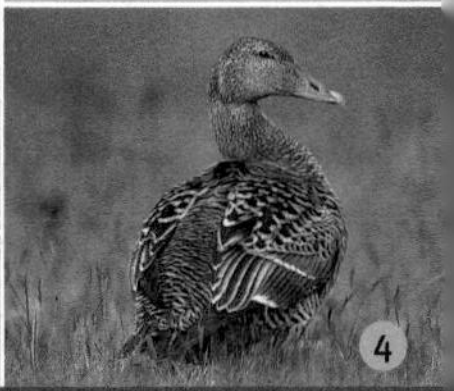
4

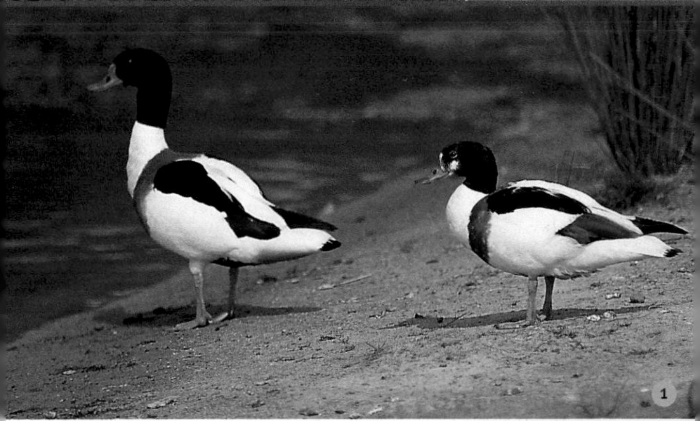
1

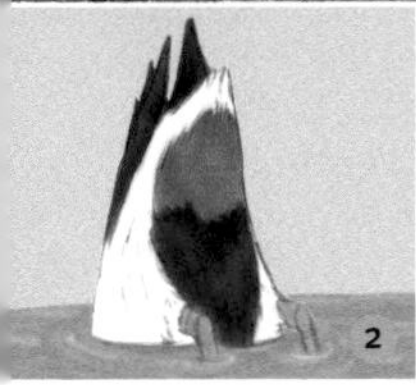
2

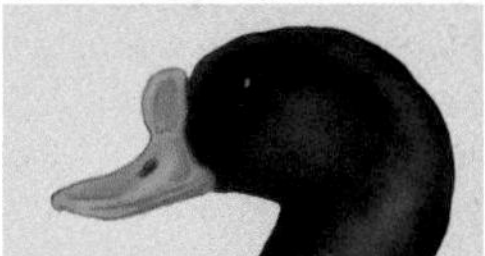
3

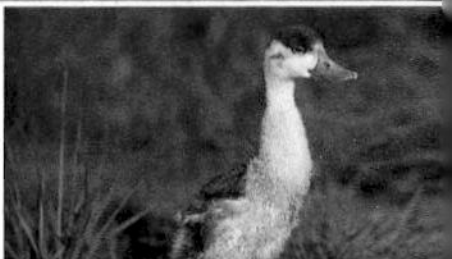
4

Greylag Goose
Anser anser

APPEARANCE: 75–90cm long; same size as Domestic Goose, but slimmer; plumage is uniform grey-brown ①; silver-grey patches on wings are visible ④; legs are flesh coloured.

VOICE: Nasal 'ga-gang-gang'; in flight a drawn out 'aahng-ong-ong'.

DISTRIBUTION: Breeds in North and East Europe; in Central Europe it exists as a half-tame park bird; today some feral birds exist on many large lakes with reed banks.

FOOD: Mainly grasses and herbaceous plants, less frequently also water plants.

TYPICAL FEATURES
Two sub-species exist in Central Europe: the eastern sub-species has a flesh-coloured bill, the western sub-species has an orange-yellow bill.

NESTING: Nests on inaccessible banks; four–eight white eggs; female incubates the eggs alone, but nidifugous young are raised by both parents.

SIMILAR SPECIES: The Bean Goose *(Anser fabalis)* ② inhabits more northern regions, but is found in Central Europe as a passage migrant and a winter visitor. Its beak has patches of black extending from the base ③.

Canada Goose
Branta canadensis

APPEARANCE: 90–110cm-long goose; back is brown, underside ranges from beige to white; neck and head are black with a bright, white, triangular patch on the cheek ①.

VOICE: Flight call is a nasal trumpeting, like a 'ga-honk', the pitch of the second syllable being higher; when seated a short 'honk'.

DISTRIBUTION: Originates from North America, but introduced to Europe as an ornamental bird; today it can be found semi-tame on park lakes, even in city centres; wild

TYPICAL FEATURES
A white 'U' divides the dark back from the black tail ②, making the Canada Goose easy to identify in flight.

populations can also be found on lakes, fish ponds and village ponds.

FOOD: Grasses and herbaceous plants, roots, seeds, water plants.

NESTING: Large, flat nest, hidden in bank vegetation or on small islands, often in small colonies; usually five–six yellowish eggs.

SIMILAR SPECIES: The Barnacle Goose *(Branta leucopsis)* ④, is a regular passage migrant and winter visitor to the North Sea coast and has a white face ③ and a grey back.

183

Mute Swan

Cygnus olor

APPEARANCE: About 1.4–1.6m long when the neck is extended; plumage is pure white ①; beak is orange-red, with a fleshy black hump ③; young birds, call cygnets, are usually grey-brown, with a grey beak.

VOICE: A snoring call, only heard during the breeding season; otherwise almost always mute; aggressive hissing when threatened.

DISTRIBUTION: A common park bird in Britain. All marked Mute Swans belong to the Queen. Today, wild populations can be found almost everywhere on lowland lakes and rivers.

TYPICAL FEATURES

A Mute Swan has a wingspan of over 2m ④. In flight, a whistling, flapping noise is heard. The Whooper Swan does not make this noise.

FOOD: Water plants, also bank vegetation; eats bread when fed.

NESTING: Nest is a large bowl constructed from reeds and bank vegetation and located near to water; five–seven grey-green to brown eggs. Young are nidifugous, but are often initially carried on the back of the adults ②.

Whooper Swan

Cygnus cygnus

APPEARANCE: The same size and colouring as the Mute Swan, but with a bright yellow beak base ①; juvenile is grey-brown ②.

VOICE: Loud trumpeting call, like a 'an-gerrh' or 'huang'; in flight, normally a three-syllable 'whoop-whoop-whoop'; in groups, also has a communication call like an 'ong' or 'geh'.

DISTRIBUTION: Breeds on lakes and river estuaries in Northern Europe and Northern Asia; a regular winter visitor, often in coastal and lowland areas.

TYPICAL FEATURES

When swimming, the neck of the Whooper Swan is not 'S'-shaped like that of the Mute Swan, but instead perfectly straight and upright ①.

FOOD: Mainly water plants, but also grass, clover, corn.

NESTING: Nest is a large bowl constructed from plant matter on the bank or in reeds; four–six cream-white eggs; young raised by both parents.

SIMILAR SPECIES: The smaller Whistling Swan (*Cygnus columbianus*) ④ has a shorter neck. The yellow patch on its beak is smaller than on the Whooper Swan and rounded at the front, not wedge-shaped ③.

185

Index

Acknowledgements

Front cover: Jay; small pictures, left to right: Great Spotted Woodpecker, Blue Tit, Hawfinch
Page 6/7: White Storks
Page 24/25: Spotted Flycatcher

Angermayer/Pölking: Page 9 top, 159 top; Angermayer/Reinhard: Page 93 bot. l., 167 top l.; Angermayer/Schmidt: Page 43 bot. r., 71 top l., 115 bot. r.; Angermayer/Ziesler: Page 113 bot. r., 147 top, 149 top r.; Danegger: Page 3 top, 6/7, 11 top, mi., 12 u., 17 top, bot. l., bot. mi.l., 18 u., 29 top l., 39 bot. r., 49 top l., 51 top l., 53 bot. l., 79 bot. r., 97 u., 115 top, 119 bot. r., 123 bot. r., 135 top, 139 bot. l., 141 top l., 143 top, bot. r., 145 bot. r., 155 bot. l., 163 top l., bot. r., 165 bot. l., bot. r., 171 top r., 179 u., bot. r., 185 top, 185 top l.; Diedrich: Page U2 r. top, 11 mi. u., 15 top r., 39 top l., 51 top r., 57 bot. r., 73 bot. r., 85 bot. r., 109 bot. r., 113 top r., 129 top r., 133 bot. l., 149 bot. r.; Fürst: Page 65 top r., 89 bot. r., 99 top; Giel: Page 173 top r.; Hüttenmoser: Page 73 top l.; Juniors/Arndt: Page 185 u.; Limbrunner: Page U2 l. top, r. mi., 2 u., 9 mi., 12 top, mi., 13 l. mi., r. top, r. mi., 14 l., 15 bot. l., bot. r., 21 l. top, l. u., r. u., 23 l., r. top, 27 top l., 35 bot. l., 41 top r., 45 top r., 51 top l., 55 top r., 57 top r., 59 top l., bot. r., 67 bot. l., 69 top, 75 u., 77 top r., 81 top r., 87 top l., top r., 89 top l., top r., 93 top l., 105 bot. r., 111 top l., 115 top r., 117 top r., 119 u., 121 top r., 125 top r., bot. r., 127 top l., bot. r., 133 bot. r., 135 top l., 137 top r., bot. r., 141 top r., bot. r., 147 bot. r., 149 top l., u., 151 (all), 153 bot. l., 155 top r., 157 u., 159 bot. r., 169 bot. r., 171 top l., bot. l., bot. r., 175 top l., bot. r., 177 top l., bot. bot. r., 181 u., bot. r., 183 top l.; Partsch: Page 13 r. u., 17 bot. mi.r., 37 bot. r., 51 u., 53 bot. r., 57 top l., 119 top r., 179 top l., 183 u., bot. r.; Pforr: 18 top, 71 top r., 111 bot. r.; Pregler: Page U1 top mi.; Reinhard: Page 8 u., 15 top l.; Schulze: Page U2 r. u., 13 l. u., 14 r., 21 l. mi., 22, 23 mi., 31 bot. l., 35 bot. r., 39 top r., 41 bot. r., 47 bot. r., 53 top r., 61 bot. r., 67 top r., 71 bot. l., 77 bot. r., 81 top l., bot. l., 83 bot. r., 87 bot. r., 93 bot. r., 95 bot. l., 107 bot. l., bot. r., 111 top r., 113 top l., bot. l., 121 top l., 123 top l., 127 top r., 139 bot. r., 145 bot. l., 159 bot. l., 183 top; Silvestris/Gross: Page 185 bot. r.; Singer: Page 117 u., 153 bot. r.; Synatschke: Page 20 u., 103 top l.; Wendl: Page U1 top l., top r., U2 l. u., 3 u., 8 top, 9 u., 11 u., 16 top, bot. mi.r., bot. r., 17 bot. r., 19 top, 24/25, 27 u., bot. r., 29 u., 31 top l., 37 u., 39 u., 41 top l., bot. l., 43 top l., 45 top l., u., 47 top l., 63 top l., top r., u., 65 bot. l., 67 top l., 69 bot. r., 73 bot. l., 75 top l., top r., 77 top l., 89 bot. l., 95 top, bot. r., 99 u., 101 top l., bot. l., bot. r., 103 bot. l., 105 top l., top r., 109 top l., 121 bot. l., 123 top r., bot. l., 125 bot. l., 131 top l., bot. l., bot. r., 135 bot. l., 137 top l., 143 u., 145 top, top l., 153 top l., 161 bot. l., 165 top l., 167 top r., 169 bot. l., 173 top l.; Wothe: Page 13 l. top, 16 bot. l., 27 top, 33 top l., bot. l., bot. r., 37 top r., 47 u., 59 u., 63 bot. r., 73 top r., 93 top l., 103 bot. r., 111 bot. l., 117 top l., 129 bot. r., 133 top, top l., 135 bot. r., 137 bot. l., 143 top l., 155 bot. r., 157 top, top l., bot. r., 161 top r., 165 top r., 167 bot. r., 175 bot. r., 181 top l.; Zeininger: Page U1 (big picture), U2 l. mi., 2 top, mi., 11 mi. top, 16 bot. mi.l., 19 u., 20 u., 21 r. top, 23 r. mi. top, r. mi. u., r. u., 29 top r., bot. r., 31 top r., bot. r., 33 top, 35 top l., bot. r., 37 top l., 43 top r., u., 47 top r., 49 top r., bot. r., bot. l., 53 top l., 55 top l., bot. l., bot. r., 59 top, 61 u., bot. r., 65 top l., bot. r., 67 top l., 69 top l., bot. l., 71 bot. r., 75 bot. r., 77 bot. l., 79 top l., top r., u., 81 bot. r., 83 top l., top r., bot. l., 85 top l., top r., bot. l., 87 u., 91 (all), 95 top l., 97 top, top l., bot. l., 99 top l., bot. r., 101 top r., 103 top, 105 bot. r., 107 top r., 109 top r., bot. l., 115 u., 119 top l., 121 bot. r., 125 top l., 127 bot. l., 129 top l., bot. l., 131 bot. r., 139 top l., top r., 141 u., 147 top l., bot. l., 153 top r., 159 top l., 161 top l., bot. r., 163 top, bot. l., 167 bot. l., 169 top l., top r., 173 u., bot. r., 175 top r., 177 top r., 179 top, 181 top r.

Acknowledgements

The dates and facts in this nature guide have been researched and checked with great care. No guarantee can, however, be given, and the publisher accepts no liability for damage to people, property or assets.

This edition first published in 2006 by New Holland Publishers (UK) Ltd
London • Cape Town • Sydney • Auckland
10 9 8 7 6 5 4 3 2 1
www.newhollandpublishers.com
Garfield House, 86–88 Edgware Road, London, W2 2EA, UK

Copyright © 2006 in translation: New Holland Publishers (UK) Ltd

ISBN 1 84537 472 X

Publishing Manager: Jo Hemmings
Senior Editor: Kate Michell
Assistant Editor: Kate Parker
Translator: American Pie, London and California

© 2005 GRÄFE UND UNZER VERLAG GmbH, Munich

Series editor: Steffen Haselbach
Editor-in-chief: Anita Zellner
Desk editors: Dr. Michael Eppinger, Dr. Helga Hofmann
Text: Dr. Helga Hofmann
Cover design: independent Medien-Design
Layout: H. Bornemann Design
Illustrations: Peter Braun, atelier amAldi
Film: Filmsatz Schröter, Munich
Production: Petra Roth
Repro: Penta, München
Printing: Appl, Wemding
Binding: Auer, Donauwörth
Printed in Germany